Tomasz Kapitaniak

Chaos for Engineers

Theory, Applications, and Control

Second, Revised Edition
With 107 Figures and Numerous Problems

Springer

Professor Tomasz Kapitaniak
Technical University of Lodz
Division of Dynamics
Stefanowskiego 1/15
90-924 Lodz
Poland
E-mail: tomaszkaocag.p.lodz.pl

Library of Congress Cataloging-in-Publication Data.

Kapitaniak, Tomasz.
Chaos for engineers : theory,applications, and control / T. Kapitaniak.-2nd rev.ed.
p. cm
Includes bibiliographical references and index.
ISBN 3540665749 (PB)
1. System engineering. 2. Chaotic behavior in systems. 3. Control theory.
4. Nonlinear theories. I. Title.
TA168.K354 2000
620'.001'171-dc21 99-053340

ISBN 3-540-66574-9 Springer-Verlag Berlin Heidelberg New York
ISBN 3-540-63515-7 1st Edition Springer-Verlag Berlin Heidelberg New York

Springer-Verlag Berlin Heidelberg New York

Cover-Design: design & production GMBH, Heidelberg

Kapitaniak: Chaos for Engineers: Theory,Applications, and Control

First Indian Reprint 2009

ISBN 978-81-8489-336-6

This edition is published by Springer (India) Private Limited, (a part of Springer Science+Business Media), Registered Office: 3rd Floor, Gandharva Mahavidyalaya, 212, Deen Dayal Upadhyaya Marg, New Delhi – 110 002, India

Printed in India by Rashtriya Printers, Delhi.

In memory of
MARTA
my grandmother

Preface

This is a revised version of the first edition, to which a new section on the estimation of the largest Lyapunov exponent using chaos synchronization has been added. This method can be very useful in engineering systems where discontinuities are common. Some new references have been added and some mistakes and misprints corrected.

I thank A. Stefański and M. Wiercigroch for helpful hints and remarks, and J. Wojewoda for his patient help with the manuscript.

Łódź, January 2000 *Tomasz Kapitaniak*

Preface

Preface to the First Edition

More than two decades of intense studies of nonlinear dynamics have shown that chaos occurs widely in engineering and natural systems; historically it has usually been regarded as a nuisance and designed out if possible. It has been noted only as irregular or unpredictable behaviour and often attributed to random external influences. Further studies showed that chaotic phenomena are competely deterministic and characteristic for typical nonlinear systems.

These studies posed questions about the practical applications of chaos. One of the possible answers is to control chaotic behaviour in such a way as to make it predictable. Recently, there have been examples of the potential usefulness of chaotic behaviour, and this has caused engineers and applied scientists to become more interested in chaos.

The main purpose of this book is to describe the new mathematical ideas in nonlinear dynamics in a way in which engineers can apply them in real physical systems. On the other hand, I tried to emphasize the importance of mathematical preciseness by giving the fundamental definitions and theorems (although without proofs).

This book is generally for readers who have the introductory background that a BSc engineering and science graduate would have; namely, ordinary differential equations and some intermediate-level dynamics and vibrations or system dynamics courses.

This book is organized as follows. In the first part (Chaps. 1–5) I describe the basic theory of nonlinear dynamics. Chap. 6 gives several engineering examples in which chaos can be expected. Finally, Chap. 7 shows how to control chaos, i.e., how to benefit from its presence.

In Chap. 1 I briefly describe the differences between linear and nonlinear oscillatory systems. The basic methods of nonlinear dynamics, such as fixed points analysis, linearization, Poincaré map, Lyapunov exponents, and spectral analysis, are introduced in Chap. 2. Chapter 3 discusses dynamics of discrete-time dynamical systems and its connection with continuous time systems. Fractals and their applications in dynamics are introduced in Chap. 4. In Chap. 5 I describe typical routes from periodic to chaotic behaviour. In Chap. 6 I present examples of chaotic behaviour of systems with direct applications in mechanical, chemical and civil engineering and in fluid dynamics. Ideas of feedback and nonfeedback methods of controlling chaos and methods of chaos synchronization are explained in Chap. 7.

In all sections a large number of examples serve to illustrate the mathematical tools. The end of an example is indicated by the following symbol . □

Theoretical chapters (1–5) finish with a number of problems. For most of the problems the assistance of a computer is necessary. A reader can either write his or her own codes or use existing software. In the second case I would recommend the *DYNAMICS* package (available with *H. Nusse and J.A. Yorke, Dynamics: numerical explorations, Springer, New York, 1994*). Alternatively one can use package *INSITE (T. Parker and L.O. Chua, Practical numerical algorithms for chaotic systems, Springer, New York, 1988)*.

Finally I would like to acknowledge the hours of discussions with S. Bishop, J. Brindley, L.O. Chua, C. Grebogi, M.S. El Naschie, T. Mullin, M. Ogorzalek, W.-H. Steeb and J. Wojewoda. Their suggestions helped to make this book clearer.

Rosanów, August 1997 *Tomasz Kapitaniak*

Contents

1. Response of a Nonlinear System

This chapter briefly describes why nonlinear phenomena are important to engineers. We show that during investigations of nonlinear systems one can observe phenomena which are not familiar from the linear theory.

Many of the oscillatory systems found in engineering problems are described by

$$m\frac{\mathrm{d}^2u}{\mathrm{d}t^2} + f\left(u, \frac{\mathrm{d}u}{\mathrm{d}t}, c, k\right) = 0. \tag{1.1}$$

A physical model for such a system is shown in Fig. 1.1. In this model m

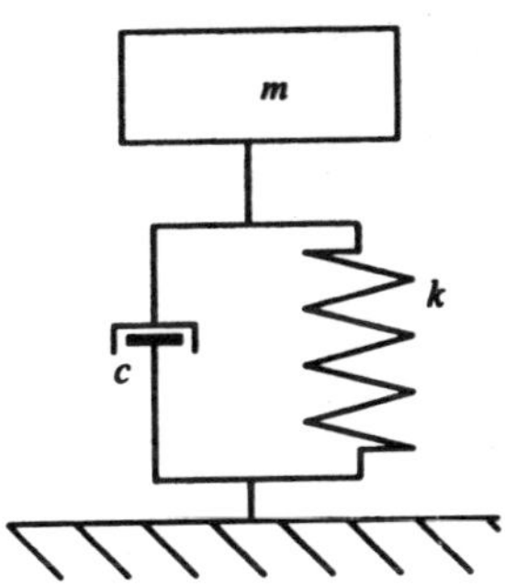

Fig. 1.1. Physical model of an oscillator

represents the mass, c is the damping coefficient and k represents the stiffness of the spring elements. The function f can be a trigonometric function (for example, a pendulum). It can also represent the nonlinear load-displacement curve of a spring (Fig. 1.2). Let us assume that we have a system without energy dissipation and that the restoring force can be approximated by

$$f(u) = ku - \frac{1}{6}\mu k u^3.$$

We arrive at the differential equation

$$\frac{\mathrm{d}^2u}{\mathrm{d}t^2} + \omega_0^2 u - \frac{1}{6}\mu\omega_0^2 u^3 = 0, \tag{1.2}$$

where $\omega^2 = k/m$. Equation (1.2) is called *Duffing's equation.* It is very important in the theory of nonlinear oscillations, because it models a large number of dynamical systems.

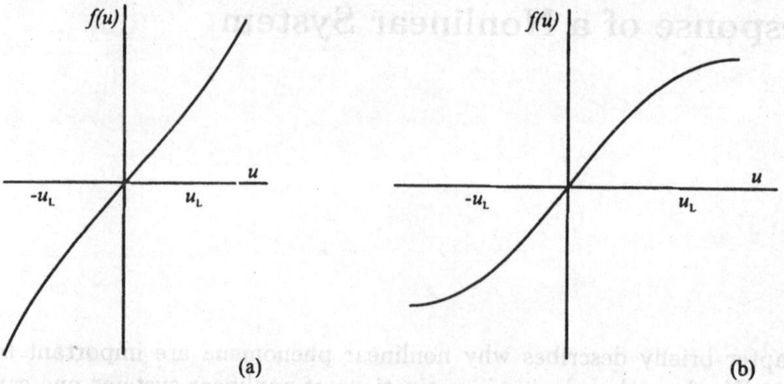

Fig. 1.2a,b. Typical characteristic of the stiffness of a spring

Let us now introduce the method of harmonic balance, which can be useful in the investigation of Duffing's equation. Let us assume that (1.2) has a solution approximated by

$$u(t) = C \sin \omega t, \tag{1.3}$$

where C is constant. Substituting (1.3) together with the identity

$$u^3 = C^3 \sin^3 \omega t \equiv C^3 \left(\frac{3}{4} \sin \omega t - \frac{1}{4} \sin 3\omega t \right)$$

into (1.2) one obtains

$$\frac{\mathrm{d}^2 u}{\mathrm{d}t^2} + \omega_0^2 \left(u - \frac{\mu x^3}{6} \right)$$

$$= \left(\omega_0^2 - \omega^2 - \mu \omega_0^2 \frac{C^2}{8} \right) C \sin \omega t + \mu \omega_0^2 \left(\frac{C^3}{24} \right) \sin 3\omega t \, ,$$

which in general is not equal to zero. One may, however, at least assure the vanishing of the factor $\sin \omega t$ by setting

$$\omega^2 = \omega_0^2 \left(1 - \mu \frac{C^2}{8} \right) \tag{1.4}$$

and assuming that C^3 is small. Equation (1.4) shows that the circular frequency of a nonlinear system $\mu \neq 0$ depends on the amplitude C. This is the characteristic property of nonlinear oscillatory systems, which does not occur in linear systems. In linear systems the frequency of oscillations depends only on the property of the system. For example, if we consider small oscillations $[u \in (-u_L, u_L)$; Fig. 1.2] we can take $\mu = 0$ and (1.4) gives

$$\omega^2 = \omega_0^2 .$$

More details on harmonic balance methods will be given in Sect. 5.5 (for a general description of this and other approximate analytical methods see

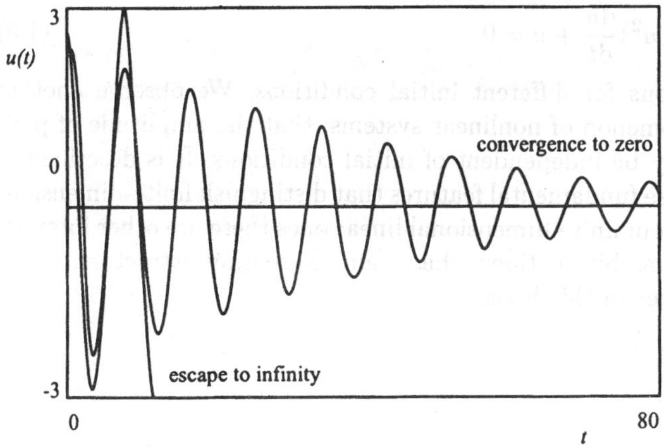

Fig. 1.3. Solutions of (1.5)

Hayashi [1.1], Nayfeh and Mook [1.2], Thomsen [1.3], Chen and Leung [1.4], and Blekhman [1.5]).

Other basic properties of nonlinear systems will be shown using the example of Van der Pol's equation

$$\frac{d^2u}{dt^2} + 0.1(1 - u^2)\frac{du}{dt} + u = 0. \tag{1.5}$$

In Fig. 1.3 typical solutions of (1.5) are shown. The two solutions shown here differ only in initial conditions and depending on them the solution can either converge to zero or escape to infinity. The property that stability may depend on initial conditions is characteristic only for nonlinear systems.

We change (1.5) to

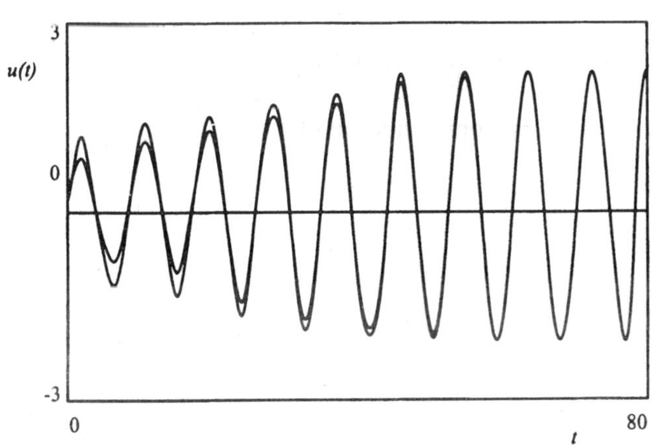

Fig. 1.4. Solutions of (1.6)

$$\frac{d^2u}{dt^2} - 0.1(1-u^2)\frac{du}{dt} + u = 0 \tag{1.6}$$

and consider solutions for different initial conditions. We observe another characteristic phenomenon of nonlinear systems: that the amplitude of periodic oscillations may be independent of initial conditions. It is described in Fig. 1.4. Besides these fundamental features that distinguish finite-dimensional nonlinear systems from finite-dimensional linear ones there are other interesting differences such as bifurcations, chaos, and coexisting attractors, which will be described later in this book.

Problems

1. Consider the linear system

$$\frac{d^2u}{dt^2} + u = 0,$$

where $u_0 = \dot{u}(0)$, $du(0)/dt = 0$. Show that the amplitude of periodic oscillations depends only on the initial conditions.

2. A mass m is suspended on a spring of stiffness k. Assume that Hooke's law applies, i.e. the rendering force is proportional to u and considering friction due to motion through the air the equation of motion is

$$m\frac{d^2u}{dt^2} + f\left(\frac{du}{dt}\right) + ku = 0.$$

Consider the solution for

$$u(0) = u_0, \qquad \frac{du(0)}{dt} = 0$$

and

$$\text{(i) } f(du/dt) = 0, \qquad \text{(ii) } f\frac{du}{dt} = 0.0001\left(\frac{du}{dt}\right)^3.$$

Can case (i) be a satisfactory approximation of case (ii)?

2. Continuous Dynamical Systems

In this chapter we describe in a simple way mathematical tools which are necessary in the analysis of dynamical systems. The fundamental notion of an attractor is introduced. We start from the fixed points, limit cycles and finally describe the properties of strange chaotic attractors. To complete this description we introduce Poincaré maps and Lyapunov exponents. Poincaré maps are tools which allow the system dimension to be reduced, an idea known to engineers from the stroboscopic lamp. Lyapunov exponents measure the divergence of trajectories starting from nearby initial conditions. These exponents are important since, in most engineering systems, initial conditions cannot be set or measured accurately. Additionally, we show that the analysis of the classical power spectrum can be also useful in analysing chaotic systems.

2.1 Phase Space and Attractors

Let us consider the first-order system of ordinary differential equations

$$\frac{\mathrm{d}\mathbf{u}}{\mathrm{d}t} = \mathbf{f}(\mathbf{u}), \qquad \mathbf{u}(t_0) = \mathbf{u}_0, \tag{2.1}$$

where $\mathbf{u} \in D \subset \mathcal{R}^n$, $t \in \mathcal{R}^+$. D is an open subset of $\mathcal{R}^n$. In most cases we have $D = \mathcal{R}^n$. A system of differential equations of the form (2.1) in which the independent variable t does not occur explicitly is called *autonomous*. Its order is equal to n.

Consider the following system of differential equations:

$$\frac{\mathrm{d}\mathbf{u}}{\mathrm{d}t} = \mathbf{f}(\mathbf{u}, t), \qquad \mathbf{u}(t_0) = \mathbf{u}_0, \qquad \mathbf{u} \in D \subset \mathcal{R}^n, \quad t \in \mathcal{R}^+. \tag{2.2}$$

If the right-hand side depends explicitly on time, (2.2) is called *nonautonomous*. If a $T > 0$ such that

$$\mathbf{f}(\mathbf{u}, t) = \mathbf{f}(\mathbf{u}, t + T)$$

exist for all $\mathbf{u}$ and t, then the equation is said to be time periodic with period T.

An nth-order time periodic nonautonomous system with period T can always be converted into an $(n+1)$th-order autonomous system by appending an extra variable

$$\Theta = \frac{2\pi t}{T}.$$

The autonomous system is given by

$$\frac{d\mathbf{u}}{dt} = \mathbf{f}(\mathbf{u}, \Theta), \qquad \mathbf{u}(t_0) = \mathbf{u}_0\,, \tag{2.3a}$$

$$\frac{d\Theta}{dt} = \frac{2\pi}{T}, \qquad \Theta(t_0) = \frac{2\pi t_0}{T}\,. \tag{2.3b}$$

The set D is called the *phase space*. In most cases we have $D = \mathcal{R}^n$. The dynamical system defined by (2.1) is the mapping

$$\Phi : \mathcal{R}^+ \times D \to \mathcal{R}^n$$

defined by the solution $\mathbf{u}(t)$. The function $\mathbf{f}$ on the right-hand side of (2.1) defines a mapping $\mathbf{f} : \mathcal{R}^n \to \mathcal{R}^n$. Thus a mapping defines a vector field on $\mathcal{R}^n$.

To show explicitly the dependence on the initial condition, the solution of (1.1) is often written as $\Phi_t(\mathbf{u}_0)$. The one-parameter family of mappings $\Phi_t : \mathcal{R}^n \to \mathcal{R}^n$ is called the flow.

The phase space of a dynamical system is a mathematical space with orthogonal coordinate directions representing each of the variables needed to specify the instantaneous state of the system. For example, the state of the particle moving in one dimension is given by its position $\mathbf{u}$ and velocity $d\mathbf{u}/dt$. Hence its phase space is two-dimensional. Generally, the dimension of the phase space of (2.1) is n (the number of the first order scalar differential equations).

Example. Consider the harmonic oscillator

$$\frac{d^2u}{dt^2} + u = 0, \qquad u_0 = u(0), \qquad \frac{du(0)}{dt} = \dot{u}_0.$$

To obtain the corresponding system of differential equations we put $u = u_1$, $du/dt = u_2$. This yields

$$\frac{du_1}{dt} = u_2, \qquad \frac{du_2}{dt} = -u_1. \tag{2.4}$$

The solution of the initial value problem is given by

$$u_1(t) = u_0 \cos t + \dot{u}_0 \sin t,$$

$$u_2(t) = -u_0 \sin t + \dot{u}_0 \cos t.$$

Thus the dynamical system is given by the mapping

$$\Phi(t, u_0, \dot{u}_0) = (u_0 \cos t + \dot{u}_0 \sin t, -u_0 \sin t + \dot{u}_0 \cos t),$$

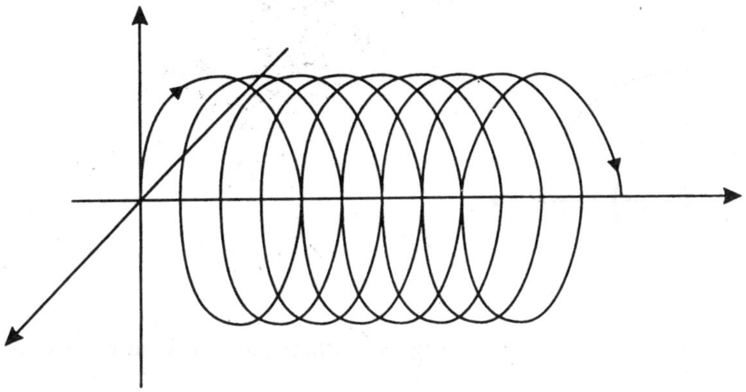

Fig. 2.1. Solution of a harmonic equation

where $\Phi : \mathcal{R}^+ \times \mathcal{R}^2 \to \mathcal{R}^2$. In the three-dimensional space $\mathcal{R}^+ \times \mathcal{R}^2$ the solution can be sketched as in Fig. 2.1.

Dividing the second equation of (2.4) by the first, we have

$$\frac{\mathrm{d}x_2}{\mathrm{d}x_1} = -\frac{x_1}{x_2} \tag{2.5}$$

and after integration

$$u_1^2 + u_2^2 = c, \tag{2.6}$$

where

$$c = u_1(0)^2 + u_2(0)^2$$

is the *constant of motion*, since

$$\frac{\mathrm{d}}{\mathrm{d}t}(u_1^2 + u_2^2) = 0.$$

Equation (2.6) describes a family of circles in the phase space $\mathcal{R}^2$ - see Fig. 2.2. □

Consider now the case of the autonomous equation (2.1) which written out in components becomes

$$\frac{\mathrm{d}u_i}{\mathrm{d}t} = f_i(\mathbf{u}), \qquad i = 1, \ldots, n\,,$$

where f_i are smooth functions. To find solutions of (2.1) one has to integrate the equation

$$\frac{\mathrm{d}u_1}{f_1} = \frac{\mathrm{d}u_2}{f_2} = \ldots = \frac{\mathrm{d}u_n}{f_n}.$$

We use one of the components of $\mathbf{u}$, say u_1 as a new independent variable. This requires $f_1(\mathbf{u}) \neq 0$. With the chain rule we obtain $(n-1)$ equations:

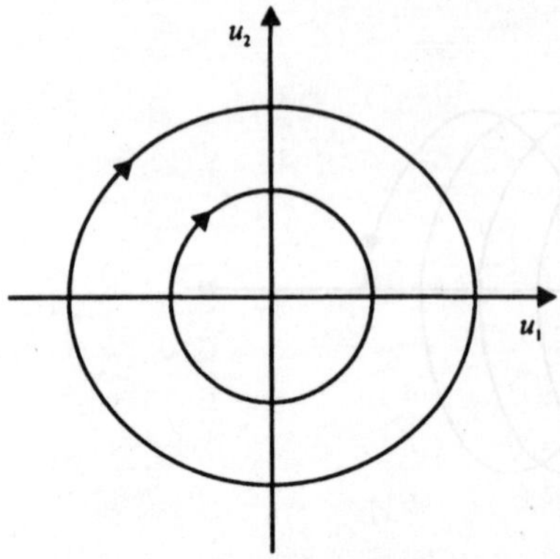

Fig. 2.2. Phase plane of a harmonic equation

$$\frac{du_2}{du_1} = \frac{f_2(\mathbf{u})}{f_1(\mathbf{u})}, \quad \ldots \quad , \frac{du_n}{du_1} = \frac{f_n(\mathbf{u})}{f_1(\mathbf{u})}. \tag{2.7}$$

Solutions of (2.7) in the phase space are called orbits. The theorem of the existence and uniqueness of the solution of an ordinary differential equation says that orbits in the phase space do not intersect. To obtain (2.7) we have assumed that $f_1(\mathbf{u}) \neq 0$. If $f_1(\mathbf{u}) = 0$ and f_2 does not vanish at these zeros, we can take x_2 as an independent variable, interchanging the roles of f_1 and f_2. If zeros of f_1 and f_2 coincide, we can take x_3 as independent variable, etc.

This construction is impossible at points $\mathbf{u}^* = (u_1^*, \ldots, u_n^*)$ where

$$f_1(\mathbf{u}^*) = f_2(\mathbf{u}^*) = \ldots = f_n(\mathbf{u}^*) = 0.$$

Definition. The point $u^* \in \mathcal{R}^n$ where

$$\mathbf{f}(\mathbf{u}^*) = \mathbf{0}$$

is called a *fixed point* of $d\mathbf{u}/dt = \mathbf{f}(\mathbf{u})$. Obviously we have

$$\Phi_t(\mathbf{u}^*) = \mathbf{u}^*.$$

Remark. Sometimes the fixed points are called equilibrium points or critical points.

Example. Consider the nonlinear differential equation

$$\frac{du}{dt} = -u^2, \qquad u_0 = u(0). \tag{2.8}$$

It is obvious that $u^* = 0$ is a fixed point, and $u(t) = 0$, $t \geq 0$ is an equilibrium solution. For $u_0 \neq 0$ the solution of the initial value problem is given by

$$u(t) = \left(\frac{1}{u_0} + t\right)^{-1}$$

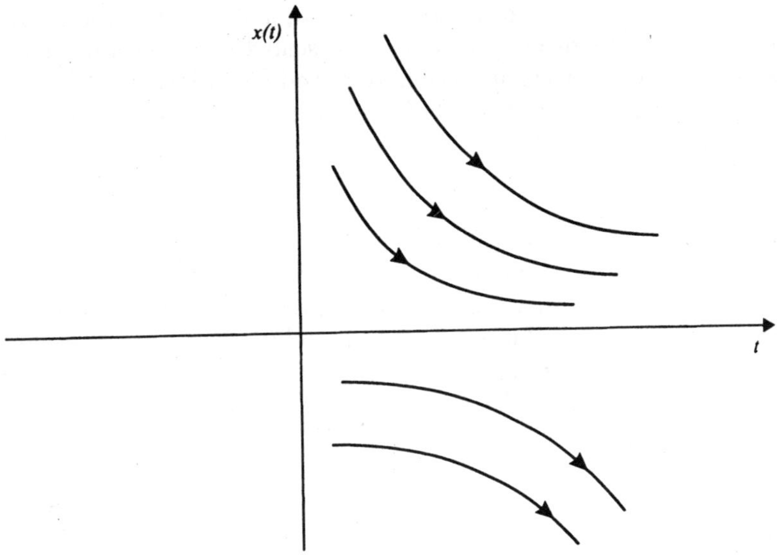

Fig. 2.3. Solution of (2.8)

and is qualitatively and quantitatively different for $u_0 > 0$ and for $u_0 < 0$ (Fig. 2.3). For $u_0 > 0$ at $t = 0$ we have

$$\lim_{t\to\infty} u(t) = 0$$

and for $u_0 < 0$ the solution becomes unbounded in a finite time. □

Example. Consider the system

$$\frac{du_1}{dt} = u_2 ,$$

$$\frac{du_2}{dt} = -u_2 - u_1 .$$

In the above example it can be easily shown that the solution tends in the limit $t \to \infty$ towards the fixed point. This phenomenon is called attraction. □

A fixed point $\mathbf{u}^*$ of the differential equation $d\mathbf{u}/dt = \mathbf{f}(\mathbf{u})$ in $\mathcal{R}^n$ is called an attractor if there exists a neighbourhood $A \subset \mathcal{R}^n$ of $\mathbf{u}^*$ such that $\mathbf{u}(t_0) \in A$ implies

$$\lim_{t\to\infty} \mathbf{u}(t) = \mathbf{u}^*.$$

If a fixed point $\mathbf{u}^* = a$ has this property for $t \to -\infty$, then $\mathbf{u}^*$ is called a *repeller*.

We have shown in the case of a harmonic equation that the solution of the differential equation $\mathrm{d}\mathbf{u}/\mathrm{d}t = \mathbf{f}(\mathbf{u})$ can be represented by the closed orbit in the phase space. Closed orbits are related to periodic solutions. If $\Phi(t)\mathbf{u}_0$ is a solution of the differential equation $\mathrm{d}\mathbf{u}/\mathrm{d}t = \mathbf{f}(\mathbf{u})$, $\mathbf{u} \in D \subset \mathcal{R}^n$ and suppose that there exists a positive number T such that

$$\Phi(t+T) = \Phi(t)$$

for all $t \in \mathcal{R}^+$, then $\Phi(t)$ is called a *periodic solution* of period T. We have the following result: A periodic solution of the autonomous equation $\mathrm{d}\mathbf{u}/\mathrm{d}t = \mathbf{f}(\mathbf{u})$ corresponds to a closed orbit in phase-space and a closed orbit corresponds to a periodic solution. A periodic solution can also be a limit cycle. If a limit cycle is reached asymptotically for $t \to \infty$ it is stable. It is an example of an attractor. If we have this property for $t \to -\infty$, a limit cycle is unstable. It is an example of a repeller.

Up till now we have introduced two examples of attractors (repellers) the critical point and the limit cycle. The attracting set can have a more complicated structure. An attractor is defined as follows:

Definition. A specific subset A of a phase space $\mathcal{R}^n$ of the differential equation $\mathrm{d}\mathbf{u}/\mathrm{d}t = \mathbf{f}(\mathbf{u})$ which is reached asymptotically as $t \to \infty$ $(t \to -\infty)$ is called an attractor (repeller).

We end this section with the famous *Poincaré-Bendixon theorem.*

Theorem. Suppose the trajectory $\Phi_t(\mathbf{u}_0)$ of the differential equation $\mathrm{d}\mathbf{u}/\mathrm{d}t = \mathbf{f}(\mathbf{u})$, $\mathbf{u} \in \mathcal{R}^2$, with flow Φ_t is contained in a bounded region D of the phase space for $t \geq 0$. Then the only possible attractors for $\Phi_t(\mathbf{u}_0)$ are a critical point or a limit cycle.

There are a number of criteria that guarantee the existence of limit cycles for certain classes of equations. Consider the second-order ordinary differential equation

$$\frac{\mathrm{d}^2u}{\mathrm{d}t^2} + f\left(u, \frac{\mathrm{d}u}{\mathrm{d}t}\right)\frac{\mathrm{d}u}{\mathrm{d}t} + g(u) = 0 .$$

Assume that the following conditions hold:

(a) $ug(u) > 0$ for all $u > 0$,

(b) $\int\limits_0^\infty g(u)\mathrm{d}u = \infty$,

(c) $f(0,0) < 0$ and there exists a $u_0 > 0$ such that $f(u, \mathrm{d}u/\mathrm{d}t) \geq 0$ for $|u| > u_0$ and every $\mathrm{d}u/\mathrm{d}t$,

(d) there exists a constant $M > 0$, such that $f(u, \mathrm{d}u/|\mathrm{d}t) \geq -M$ for $|u| < u_0$,

(e) there exists a $u_1 > u_0$ such that $\int_{u_0}^{u_1} f(u, \phi(u))du \geq 10Mu_0$,

where ϕ is an arbitrary positive and monotonically decreasing function of u. The considered equation then has at least one limit cycle.

2.2 Fixed Points and Linearisation

Let $\mathbf{u}^*$ be a fixed point of $d\mathbf{u}/dt = \mathbf{f}(\mathbf{u})$, i.e.

$$\mathbf{f}(\mathbf{u}) = \mathbf{0}.$$

In analysing fixed points we linearise the differential equation in a neighbourhood of the fixed point. Let us assume that $\mathbf{f}$ is analytic. Thus we have a Taylor series expansion of $\mathbf{f}$ around $\mathbf{u}^*$. Linearising means that we neglect higher order terms. In the case of

$$\frac{d\mathbf{u}}{dt} = \mathbf{f}(\mathbf{u})$$

we can write in the neighbourhood of the fixed point $\mathbf{u}^*$

$$\frac{d\mathbf{u}}{dt} = \frac{\partial \mathbf{f}}{\partial \mathbf{u}}(\mathbf{u} - \mathbf{u}^*) + \text{higher order terms},$$

and study the linear differential equation

$$\frac{d\mathbf{y}}{dt} = \frac{\partial \mathbf{f}}{\partial \mathbf{u}}(\mathbf{u}^*)(\mathbf{x} - \mathbf{u}^*).$$

The $n \times n$ matrix $\partial \mathbf{f}/\partial \mathbf{u}$ is called the *Jacobian matrix* or functional matrix. To simplify the notation the fixed point $\mathbf{u}^*$ is shifted to the origin of the phase space by $\bar{\mathbf{y}} = \mathbf{u} - \mathbf{a}$. Thus

$$\frac{d\bar{\mathbf{y}}}{dt} = \frac{\partial \mathbf{f}}{\partial \mathbf{u}}(\mathbf{u}^*)\bar{\mathbf{y}}.$$

We often write $\partial \mathbf{f}(\mathbf{u}^*)/\partial \mathbf{u}$ as A. It is an $n \times n$-dimensional matrix with constant coefficients. We omit the bar. So the linearised system in the neighbourhood of a fixed point $\mathbf{u}^*$ is of the form

$$\frac{d\mathbf{y}}{dt} = A\mathbf{y}. \tag{2.9}$$

Example. Consider the mathematical pendulum

$$\frac{d^2u}{dt^2} + \sin u = 0,$$

where u is the angular variable indicating the deviation from the vertical so $u \in (-\pi, \pi)$. After setting $u = u_1$, $du/dt = u_2$ one obtains

$$\frac{du_1}{dt} = u_2, \qquad \frac{du_2}{dt} = -\sin u_1 .$$

Linearisation in the neighbourhood of $(0,0)$ yields

$$\frac{du_1}{dt} = u_2, \qquad \frac{du_2}{dt} = -u_1 .$$

Linearisation in the neighbourhood of $(\pm\pi, 0)$ gives

$$\frac{du_1}{dt} = u_2, \qquad \frac{du_2}{dt} = u_1 \mp \pi .$$

Both linearisations give the results of the form of a harmonic equation. □

From further investigations of the system (2.9) we exclude the case of degenerate critical point i.e. we assume that

$$\det A \neq 0.$$

In analysing the critical points of the linear system we first determine the eigenvalues of A. The eigenvalues $\hat{\lambda}_1 \ldots, \hat{\lambda}_n$ are the solutions of the characteristic equation

$$\det(A - \hat{\lambda} I) = 0, \tag{2.10}$$

where I is the $n \times n$ identity matrix. Any vector $\mathbf{v} \neq 0$ satisfying equation

$$(A - \hat{\lambda} I)\mathbf{v} = 0 \tag{2.11}$$

is called an eigenvector.

Example. Consider the equation

$$\frac{d\mathbf{u}}{dt} = \begin{pmatrix} 1 & 1 \\ 4 & 1 \end{pmatrix} \mathbf{u} .$$

To find eigenvalues we have to consider the matrix

$$A - \hat{\lambda} I = \begin{pmatrix} 1 - \hat{\lambda} & 1 \\ 4 & 1 - \hat{\lambda} \end{pmatrix} .$$

From (2.10) one obtains

$$\det(A - \hat{\lambda} I) = (1 - \hat{\lambda})^2 - 4 = 0.$$

Thus the eigenvalues are $\hat{\lambda}_1 = -1$ and $\hat{\lambda}_2 = 3$. Let v_1 be the corresponding eigenvector for $\hat{\lambda}_1 = -1$. Then (2.11) becomes

$$\begin{pmatrix} 2 & 1 \\ 4 & 2 \end{pmatrix} \mathbf{v}_1 = 0 .$$

A non trivial solution is

$$\mathbf{v}_1 = \begin{pmatrix} 1 \\ -2 \end{pmatrix} .$$

Note that there is no unique eigenvector corresponding to $\hat{\lambda}_1 = -1$ since any multiple of $\mathbf{v}_1$ is also an eigenvector. It is left as an exercise to find the eigenvector for $\hat{\lambda}_2 = 3$ □

From linear algebra it is known that there exists a real non singular matrix T such that $T^{-1}AT$ is in the so-called normal Jordan form (for example [2.1]). If the n eigenvalues are different, then $T^{-1}AT$ is in diagonal form with the eigenvalues as diagonal elements. If there are some equal eigenvalues, the linear transformation $\mathbf{y} = T\mathbf{z}$ still leads to a simplification. One finds

$$T\frac{\mathrm{d}\mathbf{z}}{\mathrm{d}t} = AT\mathbf{z}$$

or

$$\frac{\mathrm{d}\mathbf{z}}{\mathrm{d}t} = T^{-1}AT\mathbf{z}. \tag{2.12}$$

Usually the Jordan normal form $T^{-1}AT$ is simple and we can integrate (2.12) immediately, from which $\mathbf{y} = T\mathbf{z}$ follows.

Example. Consider the linear system

$$\frac{\mathrm{d}u_1}{\mathrm{d}t} = -u_1 - 3u_2, \qquad \frac{\mathrm{d}u_2}{\mathrm{d}t} = 2u_2, \tag{2.13}$$

which can be written in the form (2.9) with the matrix

$$A = \begin{pmatrix} -1 & -3 \\ 0 & 2 \end{pmatrix}.$$

The eigenvalues of A are $\hat{\lambda}_1 = -1$ and $\hat{\lambda}_2 = 2$ and corresponding eigenvalues are given by

$$\mathbf{v}_1 = \begin{pmatrix} 1 \\ 0 \end{pmatrix}, \qquad \mathbf{v}_2 = \begin{pmatrix} -1 \\ 1 \end{pmatrix}.$$

If we take T

$$T = \begin{pmatrix} 1 & -1 \\ 0 & 1 \end{pmatrix}$$

one obtains

$$T^{-1}AT = \begin{pmatrix} -1 & 0 \\ 0 & 2 \end{pmatrix}.$$

Then under the coordinate transformation $\mathbf{z} = T^{-1}\mathbf{u}$, we obtain

$$\frac{\mathrm{d}z_1}{\mathrm{d}t} = -z_1, \qquad \frac{\mathrm{d}z_2}{\mathrm{d}t} = 2z_2,$$

which can be solved as

$$z_1(t) = c_1\mathrm{e}^{-t}, \qquad z_2(t) = c_2\mathrm{e}^{2t},$$

where c_1, c_2 depend on the initial condition. Going back to original coordinates u_1 and u_2 one obtains

$$u_1(t) = c_1 e^{-t} + c_2(e^{-t} - e^{2t}), \qquad u_2(t) = c_2 e^{2t}. \tag{2.14}$$

□

Definition. Suppose that the $n \times n$ dimensional matrix A has k negative eigenvalues $\hat{\lambda}_1, \ldots, \hat{\lambda}_k$ and $n - k$ positive eigenvalues $\hat{\lambda}_{k+1}, \ldots, \hat{\lambda}_n$ and that these eigenvalues are distinct. Let $\{\mathbf{v}_1, \ldots, \mathbf{v}_n\}$ be a corresponding set of eigenvectors. Then stable and unstable subspaces of the linear system (2.9), E^s and E^u, are the linear subspaces spanned by $\{\mathbf{v}_1, \ldots, \mathbf{v}_k\}$ and $\{\mathbf{v}_{k+1}, \ldots, \mathbf{v}_n\}$, respectively

$$E^s = \operatorname{span}\{\mathbf{v}_1, \ldots, \mathbf{v}_k\}$$
$$E^u = \operatorname{span}\{\mathbf{v}_{k+1}, \ldots, \mathbf{v}_n\}.$$

Example. Consider (2.13) from the previous example. The phase diagram can be found by sketching the solution curves defined by (2.14) (Fig. 2.4).

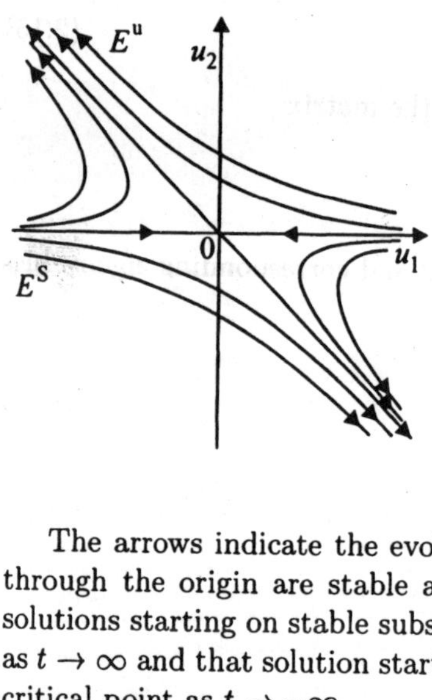

Fig. 2.4. Phase portrait of (2.13)

The arrows indicate the evolution of the system in time. The bold lines through the origin are stable and unstable subspaces of (2.13). Note that solutions starting on stable subspace E^s approach the critical point at origin as $t \to \infty$ and that solution starting an unstable subspace E^u approaches the critical point as $t \to -\infty$. □

Finally, let us consider the two-dimensional case [matrix A is 2×2 dimensional]. As the dimension of the phase-space is two, the eigenvalues $\hat{\lambda}_1$ and $\hat{\lambda}_2$ are both real or complex conjugates. The behaviour of the solution

$$\mathbf{z}(t) = \begin{pmatrix} c_1 e^{\hat{\lambda}_1 t} \\ c_2 e^{\hat{\lambda}_2 t} \end{pmatrix}$$

for $\hat{\lambda}_1 \neq \hat{\lambda}_2$ and

$$\mathbf{z}(t) = \begin{pmatrix} c_1 e^{\hat{\lambda} t} + c_2 t e^{\hat{\lambda} t} \\ c_2 e^{\hat{\lambda} t} \end{pmatrix}$$

for $\hat{\lambda}_1 = \hat{\lambda}_2 = \hat{\lambda}$ is very different depending on the choices of $\hat{\lambda}_1$ and $\hat{\lambda}_2$.

If the eigenvalues are real and have the same sign the critical point is called a node. When $\hat{\lambda}_1 < 0, \hat{\lambda}_2 < 0$ then the critical point is an attractor and if $\hat{\lambda}_1 > 0$, $\hat{\lambda}_2 > 0$ then it is a repeller (Fig. 2.5).

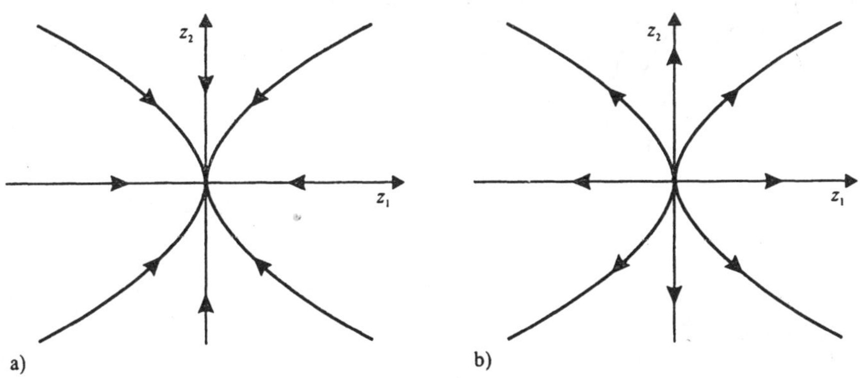

Fig. 2.5. The node. (**a**) Attractor. (**b**) Repeller

It is easy to verify that in the phase-space the orbits are straight lines through the origin when $\hat{\lambda}_1 = \hat{\lambda}_2 = \hat{\lambda}$.

If the eigenvalues are still real but have different sign, then the critical point is called a saddle. In the phase-space the orbits are given by

$$|z_1| = c|z_2|^{-|\hat{\lambda}_1/\hat{\lambda}_2|}$$

with c a constant. In this case the critical point which is called a saddle is neither an attractor nor a repeller. There exist two solutions with the property $(y_1(t), y_2(t)) \to (0,0)$ for $t \to \infty$ and two solutions with this property for $t \to -\infty$ (Fig. 2.6).

The first two of these solutions are the stable subspace of the saddle point, the other two the unstable subspace.

When the eigenvalues $\hat{\lambda}_1$ and $\hat{\lambda}_2$ are complex conjugates,

$$\hat{\lambda}_{1,2} = \mu \pm \omega i$$

with $\mu\omega \neq 0$ the complex solutions are of the form $\exp((\mu \pm \omega i)t)$. The linear combination of the complex solutions leads to real independent solutions of the form

$$e^{\mu t} \cos \omega t, \quad e^{\mu t} \sin \omega t.$$

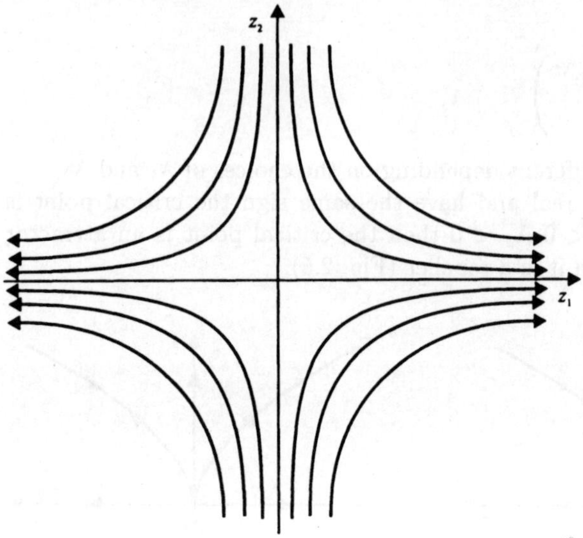

Fig. 2.6. Saddle point

The orbits spiral in or out with respect to the origin and the critical point is called a focus (Fig. 2.7).

In the case of spiralling in ($\mu < 0$) the critical point is an attractor and when $\mu > 0$ it is a repeller.

The last case is when the eigenvalues are purely imaginary. If

$$\hat{\lambda}_{1,2} = \pm \omega i,$$

then the critical point is called a centre. The solutions can be written as a combination of $\cos \omega t$ and $\sin \omega t$, the orbits in the phase-space are circles (Fig. 2.8).

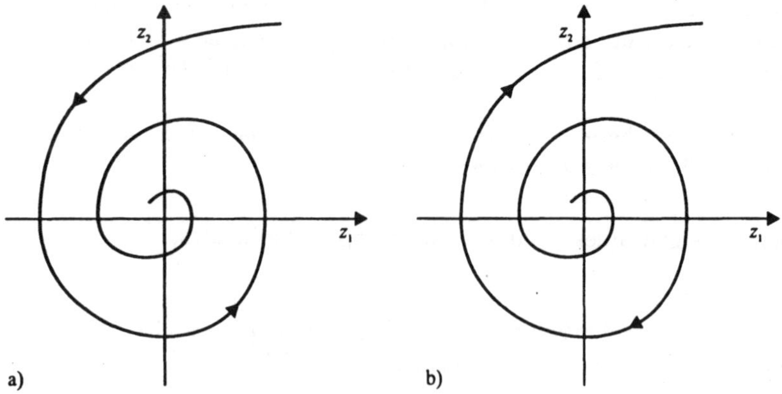

Fig. 2.7. Focus. (**a**) Attractor. (**b**) Repeller

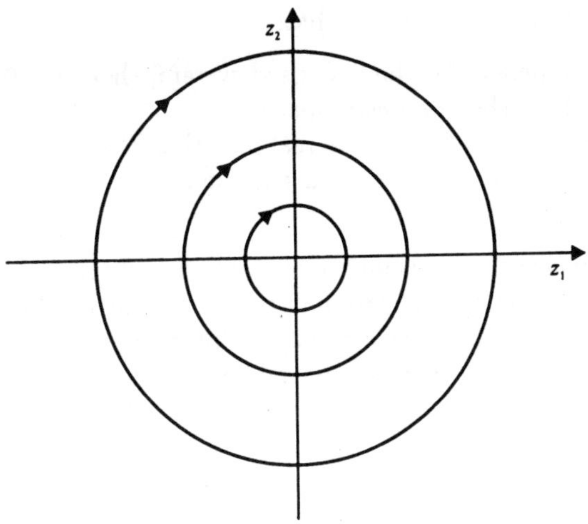

Fig. 2.8. Centre

It is clear that in this case a critical point is neither an attractor nor a repeller.

2.3 Relation Between Nonlinear and Linear Systems

The analysis of nonlinear equations usually starts with a linear analysis as described in the previous section. One tries to draw conclusions about the original nonlinear system. It turns out that some properties of the linearised system also hold for a nonlinear system. On the other hand, other properties do not carry over from linear to nonlinear systems.

We first consider $\mathrm{d}\mathbf{u}/\mathrm{d}t = \mathbf{f}(\mathbf{u})$, $\mathbf{u} \in \mathcal{R}^n$ and assume that the critical point has been transformed to $\mathbf{u}^* = 0$. Then we can write

$$\frac{\mathrm{d}\mathbf{u}}{\mathrm{d}t} = A\mathbf{u} + \mathbf{g}(\mathbf{u}) \tag{2.15}$$

with $\det A \neq 0$; A is $n \times n$ dimensional and we assume that

$$\lim_{||\mathbf{u}|| \to 0} \frac{||\mathbf{g}(\mathbf{u})||}{||\mathbf{u}||} = 0.$$

The last assumption holds under rather general conditions, for example when the function $\mathbf{f}$ is continuously differentiable in the neighbourhood of $\mathbf{u}^* = \mathbf{0}$.

In the case of $n = 2$ we have the following Poincaré result: At each singular point of the nonlinear differential $\mathrm{d}\mathbf{u}/\mathrm{d}t = \mathbf{f}(\mathbf{u})$ the classification corresponds in both type and stability with the results obtained by considering the linearised system $\mathrm{d}\mathbf{y}/\mathrm{d}t = A\mathbf{y}$, with the single exception that a centre for a linearised system may be either a centre or a focus for a nonlinear system.

For the general case $\mathbf{u} \in \mathcal{R}^n$ it can be proved that:

(a) If $\mathbf{u} = 0$ is an attractor (a repeller) for the linearised system, then $\mathbf{u} = 0$ is an attractor (or repeller) for the nonlinear equation.
(b) If the matrix A has an eigenvalue with positive real part then the critical point $\mathbf{u} = 0$ is not an attractor for the nonlinear equation.

Another important result concerns the existence of stable S and unstable U manifolds which are the nonlinear generalization of previously introduced stable and unstable subspaces (for details see Carr [2.2], Shilnikov et al. [2.3]). If Φ_t is the flow of (2.15), then S and U are positively and negatively invariant under Φ.

Definition. The set S is called a stable manifold of a critical point $\mathbf{u} = 0$ if for all initial conditions $\mathbf{u}_0 \in S$

$$\lim_{t\to\infty} \Phi_t(\mathbf{u}_0) = \mathbf{0}.$$

In the same way the set U is called an unstable manifold of a critical point $\mathbf{u} = 0$ if for all initial conditions $\mathbf{u}_0 \in U$,

$$\lim_{t\to-\infty} \Phi_t(\mathbf{u}_0) = \mathbf{0}.$$

Consider (2.15). If A has n eigenvalues with nonzero real part, $\mathbf{g}(\mathbf{u})$ is continuous and differentiable then in the neighbourhood of the critical point $x = 0$ there exist stable and unstable manifolds S and U with the same dimensions n_2 and n_u as the stable and unstable subspaces E_s and E_u of the linearised system $\mathrm{d}\mathbf{y}/\mathrm{d}t = A\mathbf{y}$. In $\mathbf{u} = \mathbf{0}$, E_s and E_u are tangent to S and U.

Example. Consider the Helmholtz oscillator

$$\frac{\mathrm{d}^2u}{\mathrm{d}t^2} + \frac{\mathrm{d}u}{\mathrm{d}t} - u - u^2 = 0.$$

Let $u_1 = u$ and $u_2 = \mathrm{d}u/\mathrm{d}t$. Then we find the two fixed points

$$(u_1^*, u_2^*) = (0,0), \qquad (u_1^*, u_2^*) = (-1,0).$$

A linear analysis shows that the first one is a saddle while the second one is a centre (Fig. 2.9). The total energy of the system H, a sum of kinetic energy $\dot{u}^2/2$ and potential energy $(-u^2/2 - u^3/3)$, is equal to

$$H = \frac{1}{2}\left(\frac{\mathrm{d}u}{\mathrm{d}t}\right)^2 - \frac{1}{2}u^2 - \frac{1}{3}u^3.$$

The solution curves in phase space are given by

$$\left(\frac{\mathrm{d}u}{\mathrm{d}t}\right)^2 - u^2 - \frac{2}{3}u^3 = C$$

(Fig. 2.9). It can be seen that the appropriate critical points of nonlinear and linearised systems correspond to each other. At the saddle point stable and unstable subspaces of a linearised system are tangent to stable and unstable manifolds of a nonlinear system.

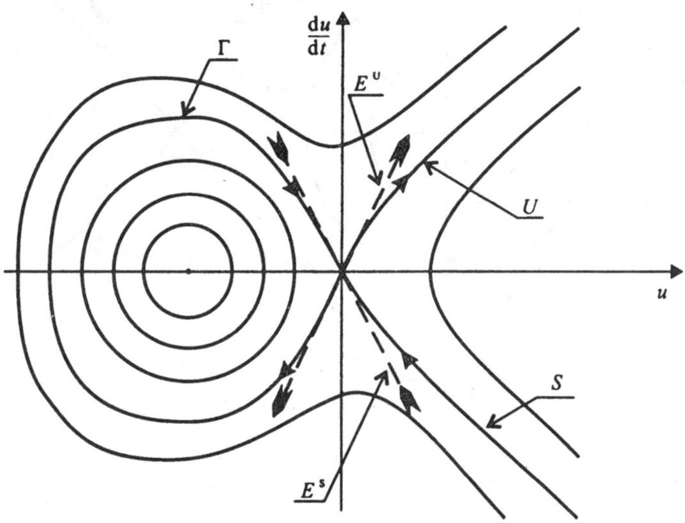

Fig. 2.9. Phase space portrait of Helmholtz oscillator

The curve $(du/dt)^2 = u^2 + 2u^3/3$, corresponding to $C \doteq 0$, goes through the point $(-3/2, 0)$ and has the saddle at the origin. This curve Γ which can be described as $\Gamma \subset S \cap U$ is called a homoclinic orbit. □

Example. Consider again a mathematical pendulum

$$\frac{d^2u}{dt^2} + \sin u = 0$$

for $u \in [-\pi, \pi]$. It can be shown that the fixed point $(0, 0)$ is a centre and fixed points $(-\pi, 0)$ and $(\pi, 0)$ are saddles. The phase space portrait of the pendulum is shown in Fig. 2.10.

Here fixed points of linear and nonlinear systems also correspond to each other. □

Consider the orbit Γ_1 which starts on an unstable manifold of a saddle at $-\pi$ and finishes at a saddle at π approaching it on the stable manifold of π. The curve Γ_2, which starts as unstable manifold of one saddle and comes to the other has a similar property.

Definition. The orbit $\gamma \subset \Gamma_1 \cup \Gamma_2$ is called a *heteroclinic orbit.*

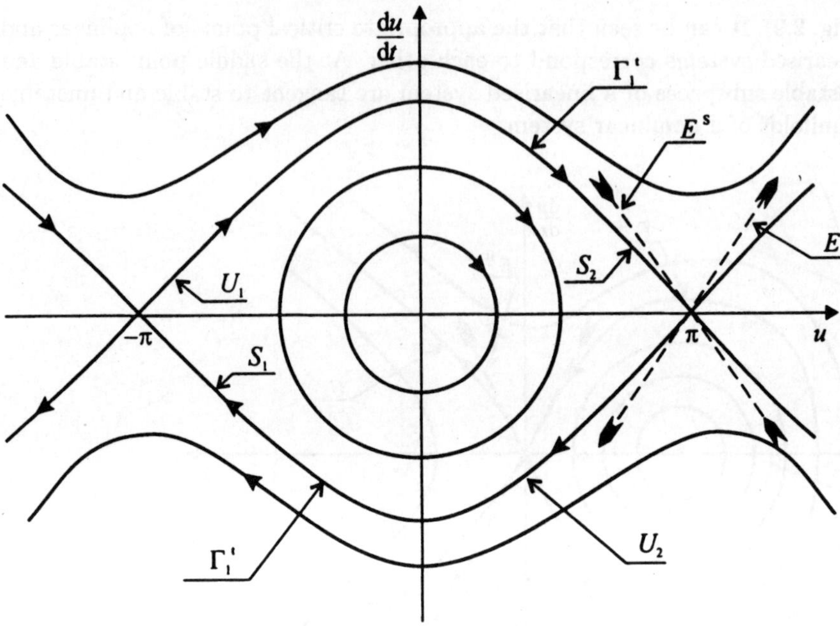

Fig. 2.10. Phase space portrait of a pendulum

2.4 Poincaré Map

A Poincaré map is a classical device for analysing dynamical systems. It is due to Henri Poincaré and his main idea to replace the flow of an nth-order continuous-time system with an $(n-1)$th-order discrete-time system. It is constructed by viewing the phase space diagram stroboscopically in such a way that the motion is observed periodically.

The definition of the Poincaré map is different for autonomous and nonautonomous system. First, let us consider an nth-order autonomous system like (2.1) and assume that it has a limit cycle Γ as shown in Fig. 2.11.

Let $\mathbf{u}^*$ be a point on the limit cycle. Let Σ be an $(n-1)$-dimensional surface transverse to Γ at $\mathbf{u}^*$. The orbit starting from $\mathbf{u}^*$ will cross Σ at x^* after a period T, of a limit cycle. Trajectories starting on Σ in a sufficiently small neighbourhood of $\mathbf{u}^*$ will, after a period T, intersect Σ in the vicinity of $\mathbf{u}^*$. Hence (2.1) and Σ define a mapping P of some neighbourhood $U \subset \Sigma$ of $\mathbf{u}^*$ onto another neighbourhood $V \subset \Sigma$ of $\mathbf{u}^*$. P is a Poincaré map of the autonomous system.

This definition of the Poincaré map is rarely used in simulations and experimental settings because it requires advanced knowledge of the position of a limit cycle. In practice, one chooses an $(n-1)$-dimensional surface Σ, which divides $\mathcal{R}^n$ into two regions. If Σ is chosen properly, then the trajectory under observation will repeatedly pass through Σ, as in Fig. 2.12. The set of these crossing points is a Poincaré map.

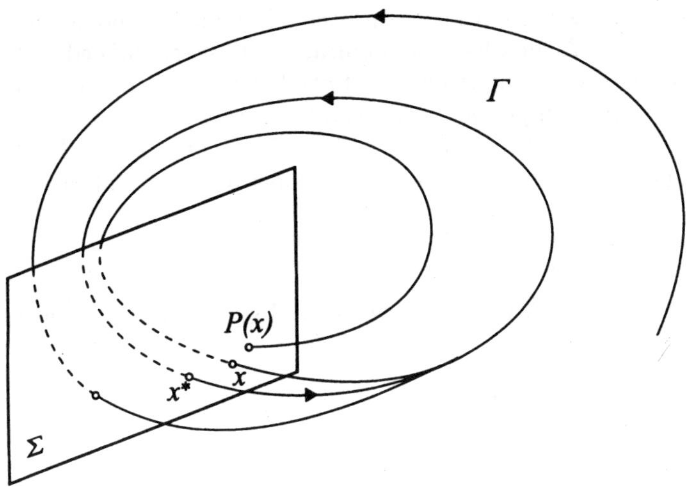

Fig. 2.11. The Poincaré map for a three-dimensional autonomous system

Example. Consider a three-dimensional autonomous system. A Poincaré map can be defined as a set

$$\sum = \{(u_1(t), u_2(t)) \quad : \quad t = t_k,\ u_3(t_k) = \text{const}\}\,.$$

Unfortunately, there is no guarantee that such a map is well defined, since $\mathbf{u}(t)$ may never intersect Σ. In the case of a system in Euclidean phase space,

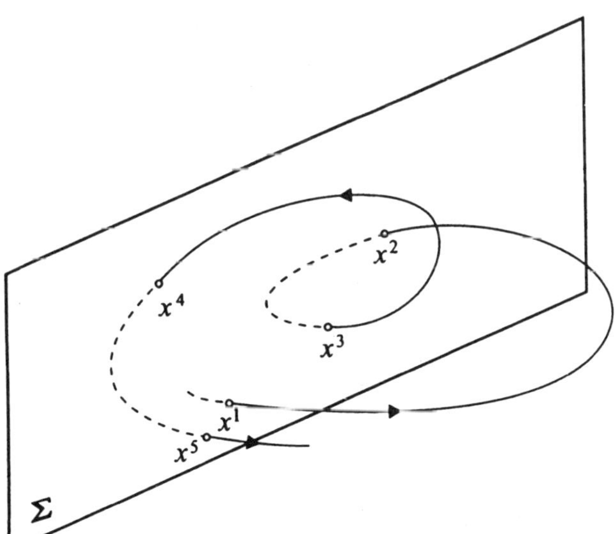

Fig. 2.12. Practical construction of the Poincaré map for an autonomous system

with bounded behaviour which does not approach an equilibrium point, there is always some choice of Σ for which the Poincaré map is well defined. □

Consider now a nonautonomous system of nth order. For a time-periodic nonautonomous system with period T that can be transformed into an $(n+1)$th order autonomous system in the cylindrical phase space $\mathcal{R}^n \times S^1$, the Poincaré map can be defined in the following way. Consider the n-dimensional surface $\Sigma \in \mathcal{R}^n \times S^1$:

$$\Sigma := \{(\mathbf{u}, \theta) \in \mathcal{R}^n \times S^1 : \theta = \theta_0\}.$$

After every period T, the orbit $\mathbf{u}(t)$ intersects Σ (Fig. 2.13). The resulting map

$$P : \Sigma \to \Sigma, \qquad (\mathcal{R}^n \to \mathcal{R}^n),$$

which maps $\mathbf{u}(t) \to \mathbf{u}(t+T)$, is a Poincaré map.

Example. Consider the forced Duffing's equation

$$\frac{\mathrm{d}^2 u}{\mathrm{d}t^2} + a\frac{\mathrm{d}u}{\mathrm{d}t} + bu + cu^3 = \beta\cos(\Omega t), \tag{2.16}$$

where a, b, c, β and Ω are constant. This equation is invariant under the following transformation

$$S : \begin{cases} (u, t) & \to (u, t + 2\pi/\Omega) \\ (\mathrm{d}u/\mathrm{d}t, t) & \to (\mathrm{d}u/\mathrm{d}t, t + 2\pi/\Omega) \end{cases}$$

and a Poincaré map can be defined as a set

$$\{(u(t), \mathrm{d}u/\mathrm{d}t) \; : \; t = t_0 + 2k\pi/\Omega, \; k = 1, 2, \ldots\}.$$

□

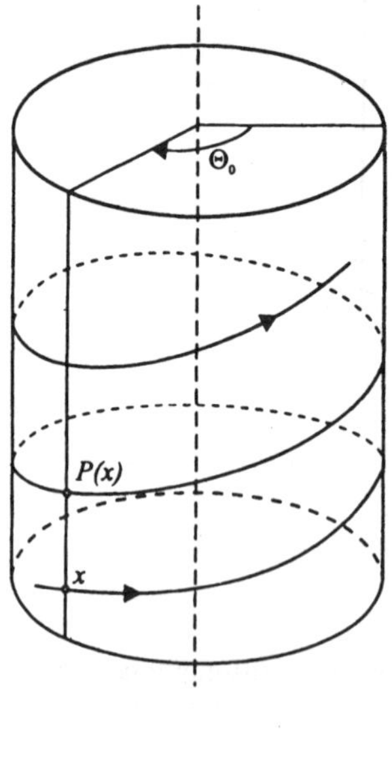

Fig. 2.13. The Poincaré map of a one-dimensional nonautonomous system

Unfortunately, there is no general analytical method of computing a Poincaré map associated with an arbitrary ordinary differential equation. The widespread use of computers with graphic facilities to examine dynamical systems has led to the method of Poincaré maps being one of the most popular and illustrative methods in these investigations.

2.5 Lyapunov Exponents and Chaos

Lyapunov exponents, named after the Russian mathematician A.M. Lyapunov, can be used to obtain a measure of the sensitive dependence of the solution of

$$\frac{d\mathbf{u}}{dt} = \mathbf{f}(\mathbf{u}), \qquad \mathbf{u} \in D \subset \mathcal{R}^n,$$

on the initial condition. Lyapunov exponents are a generalization of the eigenvalues of a fixed point. The linearised equation is given by

$$\frac{d\mathbf{y}}{dt} = \frac{\partial \mathbf{f}}{\partial \mathbf{u}}(\Phi(\mathbf{u}_0))\mathbf{y},$$

where $\mathbf{u}_0 = \mathbf{u}(t = 0)$. The solution of this system can be written as

$$\mathbf{y}(t) = U^t_{\mathbf{u}_0}\mathbf{y}_0,$$

where $U^t_{\mathbf{u}_0}$ is the fundamental matrix of the linearised equation. The fundamental matrix satisfies the chain rule

$$U^{t+s}_{\mathbf{u}_0} = U^t_{\mathbf{u}_0} \circ U^s_{\mathbf{u}_0}.$$

The asymptotic behaviour of the fundamental matrix for $t \to \infty$ can be characterised by the following exponents:

$$\lambda(V^k, \mathbf{u}_0) = \lim_{t\to\infty} \ln \frac{U^t_{\mathbf{u}_0}\mathbf{e_1} \wedge U^t_{\mathbf{u}_0}\mathbf{e_2} \wedge ... \wedge U^t_{\mathbf{u}_0}\mathbf{e_n}}{||\mathbf{e}_1 \wedge \mathbf{e}_2 \wedge \ldots \wedge \mathbf{e}_k||}.$$

Let $m_1(t), \ldots, m_n(t)$ be the eigenvalues of the solution of

$$\frac{d\mathbf{y}}{dt} = A(x_0)\mathbf{y},$$

$$\mathbf{y}(t) = e^{A(x_0)t},$$

where

$$A(x_0) = \frac{\partial \mathbf{f}}{\partial \mathbf{u}}(x_0).$$

The Lyapunov exponents of x_0 are

$$\lambda_i(x_0) = \lim_{t\to\infty} \frac{1}{t} \ln |m_i(t)|$$

whenever the limit exists.

Example. Consider the Lyapunov exponents of the fixed point $\mathbf{u}^*$ of the dynamical system $\mathrm{d}u/\mathrm{d}t = f(u)$. Let

$$\hat{\lambda}_1, \ldots, \hat{\lambda}_n$$

be the eigenvalues of the linearised equation $\mathrm{d}u/\mathrm{d}t = A(\mathbf{u}^*)$, then $m_i(t) = \mathrm{e}^{\hat{\lambda}_i t}$, and

$$\lambda_i = \lim_{t\to\infty} \frac{1}{t} \ln |\mathrm{e}^{\hat{\lambda}_i t}| = \lim_{t\to\infty} \frac{1}{t} \operatorname{Re} [\hat{\lambda}_i]t = \operatorname{Re}[\hat{\lambda}_i]. \qquad \square$$

As shown in the above example, the Lyapunov exponents are equal to the real parts of the eigenvalues of the critical point. They indicate the rate of contraction (when $\lambda_i < 0$) or expansion (when $\lambda_i > 0$) close to the fixed point. The subspaces in which the expansion or contraction occurs are determined by the appropriate eigenvectors of $A(a)$.

Positive one-dimensional Lyapunov exponents mean that two nearby trajectories (trajectories for slightly different initial conditions) diverge exponentially.

Direct application of the definition of Lyapunov exponents to their calculation is based on the integration of a linearised equation

$$\frac{\mathrm{d}\mathbf{y}}{\mathrm{d}t} = A(x_0)\mathbf{y} \tag{2.17}$$

for $\mathbf{y}(0) = (1, 1, \ldots, 1)$ together with the equations of motion. If the integration takes place for a sufficiently long time T, then

$$\lambda_i = \frac{1}{T} \ln |m_i(T)|.$$

This approach is not very workable since, when at least one Lyapunov exponent is positive, $\mathbf{y}(t)$ is unbounded as $t \to \infty$ and serious numerical problems can arise in the integration of (2.18).

A more suitable method takes advantage of the fact that almost every initial perturbation of the system grows exponentially on the average. Here we describe how to compute the largest one-dimensional Lyapunov exponent. We choose an initial condition x_0 and an initial perturbation δx_0. Let

$$y^{(0)} = y_0, \qquad u^{(0)} = \frac{y_0}{||\delta y_0||},$$

and $y^{(0)} = y_0$. We integrate the linearised equation from $u^{(0)}$ for time T. We obtain

$$y^{(1)} = y(T, u^{(0)}, y^{(0)}) = y(u^{(0)}, T)u^{(0)}.$$

Let

$$u^{(1)} = \frac{y^{(1)}}{||\delta y^{(1)}||}$$

be the normalised version of $\delta y^{(1)}$. We integrate the linearised equation from $u^{(1)}$ for T to obtain

$$y^{(2)} = \delta(T, u^{(1)}, y^{(1)}) = y(y^{(1)}, T)u^{(1)},$$

where $y^{(1)} = y(x^{(0)}, T)$. Repeating this procedure K times one obtains

$$y(KT, y_0, \mathbf{u}_0) = ||y^{(K)}|| \dots ||y^{(1)}||u^{(K)}$$

and for K large enough,

$$\lambda_1 \approx \frac{1}{KT} \ln ||y(kT, \mathbf{y}_0, \mathbf{u}_0)|| = \frac{1}{kT} \ln \prod_{k=1}^{K} ||\mathbf{y}^{(k)}|| = \frac{1}{kT} \sum_{k=1}^{K} \ln ||\mathbf{y}^{(k)}||.$$

Typically T is chosen to be ten to twenty times the natural period of the system. Too small or too large T results in serious numerical difficulties. In a similar way it is possible to calculate all n Lyapunov exponents. The algorithm for calculating all Lyapunov exponents is somewhat beyond the scope of this text. Details can be found in Wolf et al. [2.4], Parker and Chua [2.5], Kapitaniak [2.6].

Exploiting the fact that Lyapunov exponents measure the rate of contraction or expansion they can be used as a simple criterion to distinguish between conservative and dissipative systems. For

$$\sum_{i=1}^{n} \lambda_i = 0$$

the volume of a solution in phase space is conserved and in this case we have a conservative system. In dissipative systems, the phase space is contracted with

$$\sum_{i=1}^{n} \lambda_i < 0.$$

Note that a dynamical system has an attractor only when

$$\sum_{i=1}^{n} \lambda_i \leq 0$$

since, for

$$\sum_{i=1}^{n} \lambda_i > 0,$$

the system is expanding and may never reach any attractor.

Lyapunov exponents help in the characterization of different types of attractors of dynamical systems. If we have a one-dimensional dissipative system, the only possible attractor $\lambda_1 = (-)$ ($-$ represents the sign of λ_i) is represented by a (fixed) point in phase space.

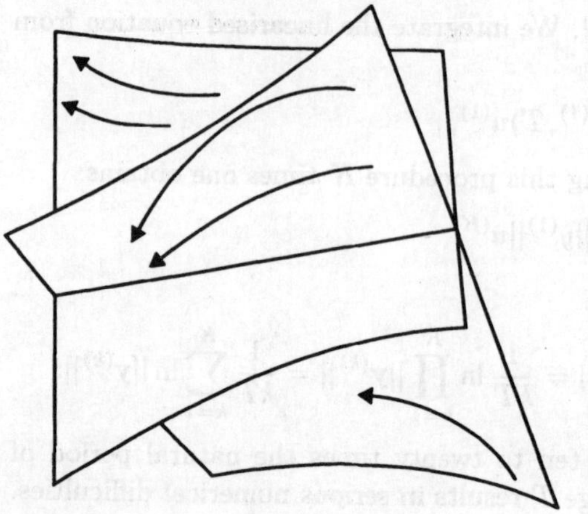

Fig. 2.14. Exponential spreading and contraction of an attractor with $(+,0,-)$ Lyapunov exponents

For dissipative system in two-dimensional phase space, one has to consider two Lyapunov exponents. The combination $(\lambda_1, \lambda_2) = (-,-)$ provides a point-like attractor and the combination $(\lambda_1, \lambda_2) = (0,-)$ is represented by a limit cycle in the phase space, which corresponds to the periodic solution.

For a three-dimensional dissipative system there are three possible stable types of solutions $(-,-,-)$ point-like attractor, $(0,-,-)$ limit cycle, and $(0,0,-)$ focus two-frequency quasi-periodicity. In addition, there are non-trivial attractors with $(+,0,-)$ whenever

$$\sum_{i=1}^{n} \lambda_i < 0.$$

Attractors with positive Lyapunov exponents are called strange chaotic attractors, and the solution of (2.1) is called chaotic if at least one one-dimensional Lyapunov exponent is positive.

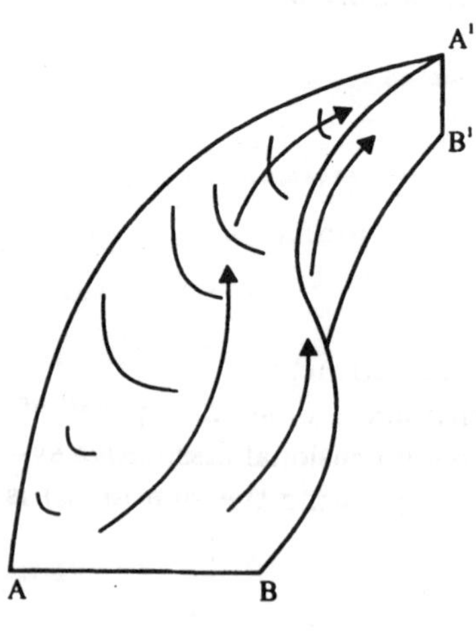

Fig. 2.15. Folding of sheets necessary to obtain exponential divergence of nearby trajectories

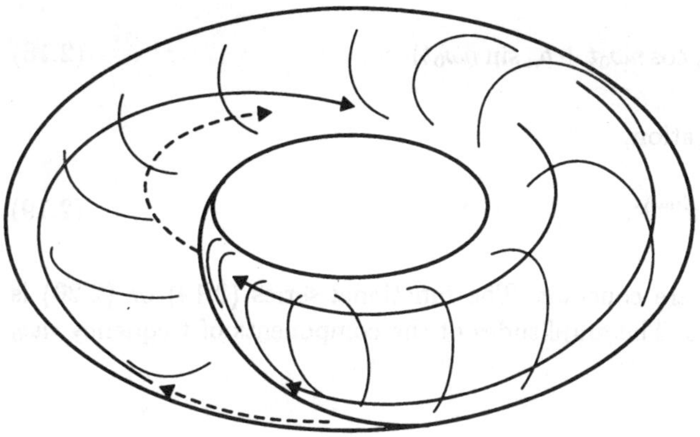

Fig. 2.16. Attractor obtained after connection $A \to A^1$ and $B \to B^1$

In a strange chaotic attractor the positive Lyapunov exponent indicates exponential spreading within the attractor in the direction transverse to the flow and the negative exponent indicates exponential contraction onto the attractor. Under the action of such a flow, phase space volumes evolve into sheets, as shown in Fig. 2.14.

Exponential divergence of nearby trajectories within a compact subspace requires the folding of sheets. A simple example of this process is shown in Fig. 2.15. Trajectories diverge exponentially within a sheet, then the sheet folds and connects back to itself (A is connected with A^1, and B with B^1), finding an attractor shown in Fig. 2.16. Generally, this type of an attractor is not simply a sheet with a single fold, but a sheet folded and refolded infinitely by a flow.

2.6 Spectral Analysis

The time evolution of a dynamical system is represented by the time variation $f(t)$ (time series) of its dynamical variables. If the time dependent function $f(t)$ fulfills certain conditions given later it can be represented as a superposition of periodic components. The determination of these components is called spectral analysis.

If the function $f(t)$ is continuous and its derivative $\mathrm{d}f(t)/\mathrm{d}t$ is continuous, $f(t)$ is periodic, i.e.

$$f(t) = f(t + nT)$$

with n being a positive or a negative integer and T being the basic periodicity, then $f(t)$ can be expressed as a linear combination of oscillations whose frequencies are integer multiples of a basic frequency ω_0, i.e.

$$f(t) = \sum_{n=-\infty}^{\infty} (a_n \cos n\omega_0 t + b_n \sin n\omega_0 t) \tag{2.18}$$

or, using complex notation,

$$f(t) = \sum_{n=-\infty}^{\infty} c_n e^{in\omega_0 t}, \tag{2.19}$$

where a_n, b_n and c_n are constant. The functional series (2.19) or (2.20) is called a *Fourier series.* The amplitudes of the components of frequency $n\omega_0$ are given by

$$a_n = \frac{\omega_0}{\pi} \int_{-\pi/\omega_0}^{\pi/\omega_0} f(t) \cos(n\omega_0 t) dt, \tag{2.20}$$

$$b_n = \frac{\omega_0}{\pi} \int_{-\pi/\omega_0}^{\pi/\omega_0} f(t) \sin(n\omega_0 t) dt, \tag{2.21}$$

or

$$c_n = \frac{\omega_0}{\pi} \int_{-\pi/\omega_0}^{\pi/\omega_0} f(t) e^{-in\omega_0 t} dt.$$

For more details on spectral analysis and conditions under which a more general class of functions f can be represented by Fourier series, see for example Kaplan [2.7].

When f is not periodic it can be expressed in terms of oscillations with continuum frequencies. Such a representation is called the Fourier transform of f. In this case the spacing between the frequency components becomes infinitesimal and the discrete spectrum of frequency components becomes continuous.

Consider the following transformations

$$T \to \infty, \qquad n\omega_0 \to \omega,$$

where ω is a continuous variable, and

$$a_n \to a(\omega) d\omega.$$

This transformation allows the transition from the Fourier series to the Fourier transform. The appropriate limits lead to the Fourier transform $a(\omega)$ of the function $f(t)$

$$a(\omega) = \frac{1}{2\pi} \int_{-\infty}^{\infty} f(t) e^{i\omega t} dt. \tag{2.22}$$

The inverse transform is given by

$$f(t) = \int_{-\infty}^{\infty} a(\omega) e^{-i\omega t} d\omega.$$

The Fourier transform $a(\omega)$ can be complex. Therefore it is useful to define a real-valued function:

$$S(\omega) = |a(\omega)|^2,$$

which is called the *power spectrum*. The power spectrum is the quantity which is very useful in a number of practical applications, for example it allows to determine the main frequencies of the considered system and to avoid resonances in experiments. This quantity is typically calculated in experimental and numerical works.

Example. Let $t \in (0, 2\pi)$. Consider the function

$$f(t) = \frac{\pi - t}{2},$$

which generally is not periodic. As we are interested in its Fourier representation in the interval $(0, 2\pi)$ of $f(t)$, let us consider the function $f(t)$ shown in Fig. 2.17 which is periodic with $T = 2\pi$ and, for $t \in (0, 2\pi)$, $f(t) = \tilde{f}(t)$.

Applying (2.19) one obtains

$$\frac{\pi - x}{2} = \sum_{h=1}^{\infty} \frac{\sin nx}{2} \quad \text{for} \quad 0 < x < 2\pi. \tag{2.23}$$

The function $\tilde{f}$ has the form of a so-called 'sawtooth' function and has applications in electronics. □

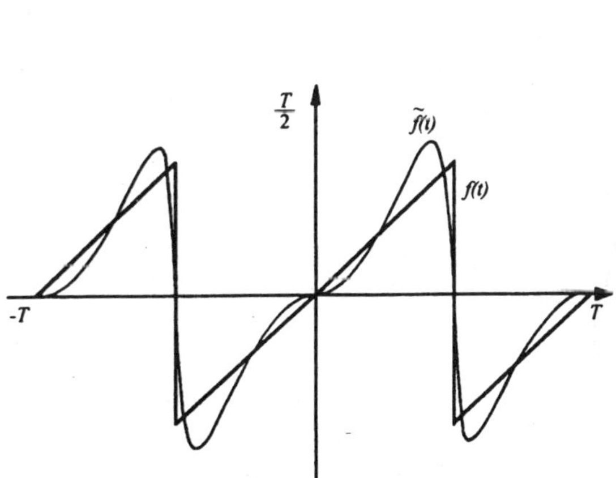

Fig. 2.17. Sawtooth function $\tilde{f}(t)$

Example. Let us consider the solution of the damped oscillator

$$\frac{d^2u}{dt^2} + c\frac{du}{dt} + \omega_0^2 u = 0,$$

which is given by

$$u(t) = Ae^{-ct}e^{i\omega_0 t},$$

where c is the damping coefficient, ω_0 is the frequency of undamped oscillations and A is a constant which depends on initial conditions. The application of (2.22) allows the calculation of the integral for $a(\omega)$:

$$a(\omega) = \frac{A}{2\pi[c - i(\omega - \omega_0)]}$$

and the power spectrum is as follows

$$S(\omega) = \frac{1}{4\pi^2[c^2 + (\omega - \omega_0)^2]},$$

The power spectrum $S(\omega)$ is shown in Fig. 2.18. $S(\omega)$ is symmetric about the dominant frequency ω_0 which is the natural frequency of the undamped system, but as a damping parameter is positive it has a finite width γ. If $c \to 0$, then $S(\omega)$ becomes very sharp. For $c = 0$ (undamped oscillator) the solution is periodic: $u(t) = Ae^{i\omega_0 t}$. □

The analytical calculation of the Fourier transform is quite complicated for practical systems but numerical methods are straightforward. When the data are discrete or digitized it is possible to adopt a method which does not involve numerical integration. This algorithm is called the fast Fourier transform (FFT). It takes advantage of certain symmetry properties in the trigonometric functions at their points of evaluation in order to gain in speed over more conventional methods. The programme for computing FFT can be found in most software libraries.

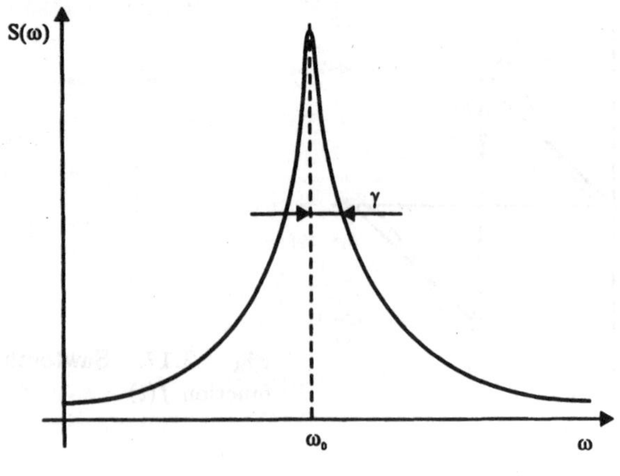

Fig. 2.18. Power spectrum of the response of a damped linear oscillator

2.7 Description of Different Attractors

In this section we try to give the characteristics of different types of attractors using the description introduced in the previous section. Let us consider the nonlinear oscillator

$$\frac{d^2u}{dt^2} + 0.1\frac{du}{dt} + \frac{1}{2}(u^2 - 1)u = 0.17\cos \Omega t \tag{2.24}$$

for $\Omega = 0.37,\ 0.38,\ 0.40$. Different types of attractors are shown in Fig. 2.19. In Fig. 2.19a,b the response of (2.24) is periodic with a period of excitation force $T = 2\pi/\Omega$. Depending on the initial condition the two different attractors shown in Fig. 2.19a,b are possible. The phase-space portrait is characterized by a simple closed orbit while a Poincaré map displays one point. The period of the solution is visible from the power spectrum which consists of a single component of the frequency Ω. Fig. 2.19c shows the response of (2.24) which is periodic but with period $3T$ (three times period of the excitation force). Here we can see a phase-space portrait with a more complicated but still closed orbit and three points in a Poincaré map. In power spectra the frequency $\Omega/3$ is visible. Fig. 2.19d presents the chaotic response of (2.24). The phase-space portrait shows the complicated unclosed curve and the Poincaré map consists of an infinite number of points. As the response is not periodic, the power spectrum must be expressed in terms of oscillations with a continuum of frequencies.

Note that intersections of the orbits visible in phase-space portraits are due to the fact that the phase space is three-dimensional $(x, \dot{x}, \Omega t)$ and the figures show the projections of the trajectories. Of course, there are no intersections in three-dimensional phase space.

The described properties of phase-space portraits, Poincaré maps and power spectra of different types of solution are general with the exception of the properties of Poincaré maps for periodic and quasiperiodic solutions of an autonomous system. For a nonautonomous system the k-periodic solution corresponds to k points in a Poincaré map. For an autonomous system there is not such a relation as the number of points on a Poincaré map depends on the selection of cross-section Σ_j as described in Fig. 2.20. The properties of the different types of attractors are summarised in Table 2.1.

2.8 Reconstruction of Attractor from Time Series

If the dynamical system can be modeled by systems of ordinary differential equations, then one can make an unambiguous identification between the state variables and the phase-space coordinates of our dynamical systems approach. With these coordinates, when the derivatives

$$\frac{\delta f_i(\mathbf{u}(t))}{du_j}$$

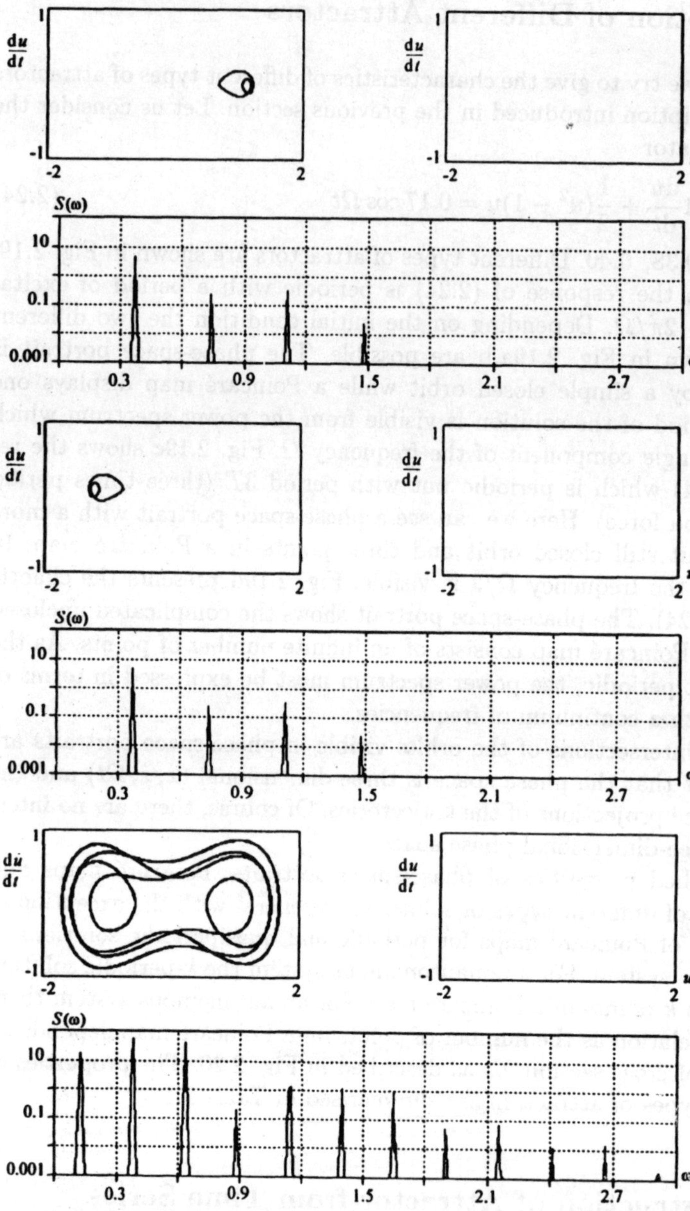

Fig. 2.19a–d. Phase space portraits, Poincaré maps and power spectra of different types of solution of (2.24). (a), (b): Period-T. (c): Period-$3T$. (d) Chaotic

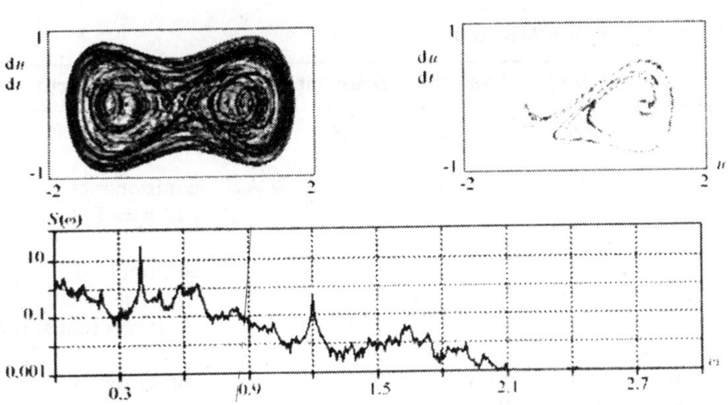

Fig. 2.19a–d. (continued)

exist, as has been shown in the previous section, the determination of the attractor properties is straightforward. An experimental system for which the equations of motion are unknown presents greater difficulties. In this case the attractor has to be reconstructed from the measured time series $z(t)$. The idea, which is justified by embedding theorems [2.8], [2.9], is as follows: For almost every observable $z(t)$ and time delay τ an m-dimensional portrait constructed from the vectors

$$[z(t_0), z(t_0+\tau), ..., z(t_0+(m-1)\tau)]$$

can have the same properties (the same Lyapunov exponents) as an original attractor. The graphical description of this idea for *embedding dimension* $m = 3$ is shown in Fig. 2.21.

In Fig. 2.22 we can compare the projections of the attractors of the Rössler system

$$\frac{dx}{dt} = -y - z$$
$$\frac{dy}{dt} = x + 0.2y \tag{2.25}$$
$$\frac{dz}{dt} = 0.2 + xz - 5.7z\ .$$

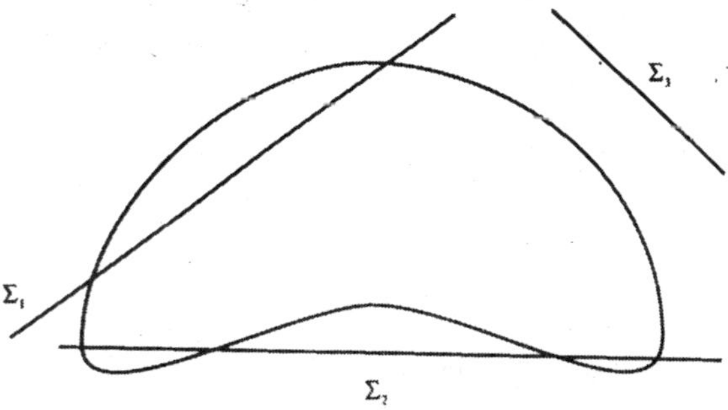

Fig. 2.20. Different selections of Poincaré map for two-dimensional autonomous system

Table 2.1. Different types of attractors

Attractor	Poincaré Map	Lyapunov exponents	Power spectrum
equilibrium	does not exist	$0 > \lambda_1 \geq \ldots \geq \lambda_n$	does not exist
periodic (limit cycle)	one or more points	$\lambda_1 = 0, 0 > \lambda_1 \geq \ldots \geq \lambda_n$	components $n\Omega\ n = 1, 2\ldots$
2-periodic	one or more closed curves	$\lambda_1 = \lambda_2 = 0$ $0 > \lambda_3 \geq \ldots \geq \lambda_n$	components $n\Omega_{1,2}\ n = 1, 2$ Ω_i incommensurate
k-periodic (k-torus)	one or more ($k-1$)-tori	$\lambda_1 = \ldots = \lambda_k = 0$ $0 > \lambda_{k+1} \geq \cdots \geq \lambda_n$	components $n\Omega_i,\ n = 1, 2\ldots$ Ω_i incommensurate
strange chaotic	continuum points	$\lambda_1 > 0$ $\sum_{i=1}^{n} \lambda_i < 0$	continuum number of components

In Fig. 2.22a we see the attractor obtained from the direct integration of (2.25), while Fig. 2.22b shows the attractor reconstructed from time series.

Strictly speaking, the phase portrait obtained by this procedure gives an embedding of the original manifold. The choice of the time delay τ is almost but not completely arbitrary. If we have a system modeled by partial differential equations we do not know how to choose embedding dimension m. Generally, if it is possible to follow N independent variables, m must satisfy the inequality

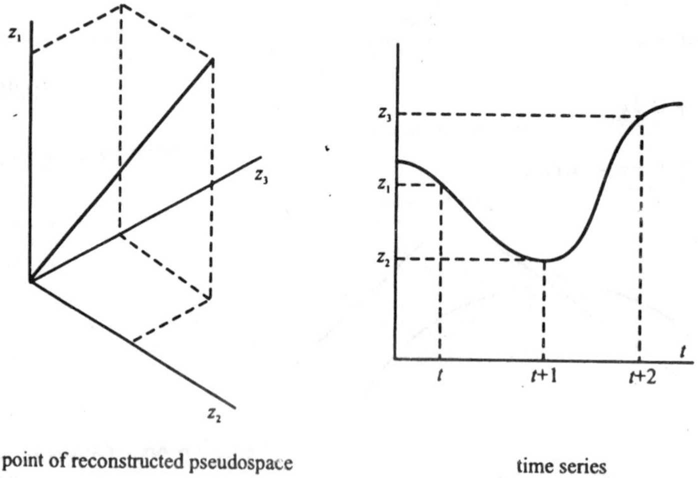

Fig. 2.21. Reconstruction of attractor from time series

(a)

(b)

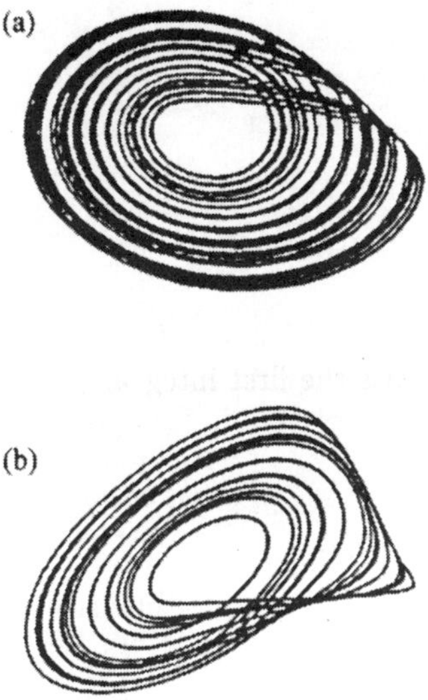

Fig. 2.22. Rössler attractor. (**a**) Original. (**b**) Reconstructed from time series $x(t)$

$$m > 2N + 1.$$

However, in most cases an unambiguous phase portrait can be obtained with many fewer dimensions than the inequality requires. In practice m is increased by one at a time until the correlation dimension (see Sect. 4.2) ceases to increase.

Problems

1. Consider the linear differential equation

$$\frac{d^2u}{dt^2} - \lambda\frac{du}{dt} - (\lambda - 1)(\lambda - 3)u = 0,$$

where λ is a real parameter. Find the fixed points and characterize them.

2. The Volterra–Lotka equations are given by

$$\frac{du_1}{dt} = au - bu_1u_2,$$

$$\frac{du_2}{dt} = bu_1u_2 - cu_2$$

with $u_1, u_2 \geq 0$ and a, b, c positive constants describe the interaction of two species, where u_1 denotes the population density of the prey, x_2 the population density of the predator. The survival of the predators depends completely on the presence of prey, so if $u_1(0) = 0$ then we have

$$\frac{du_2}{dt} = -cu_2.$$

It follows that

$$u_2(t) = u_2(0)e^{-ct}$$

and

$$\lim_{t\to\infty} u_2(t) = 0.$$

Show that the Volterra–Lotka equation admits the first integral

$$I(u_1, u_2) = u_1 u_2 e^{u_1+u_2}.$$

3. Consider the linear equation

$$\frac{d\mathbf{u}}{dt} = A\mathbf{u}, \qquad \mathbf{u} \in D \subset \mathcal{R}^n.$$

Let $\delta = \det A$ and $\tau = \text{tr}A$. Show that
(a) if $\delta < 0$ then there is a saddle at the origin;
(b) if $\delta > 0$ and $\tau^2 - 4\delta \geq 0$ then there is node at the origin (show when it is an attractor);
(c) if $\delta > 0$, $\tau^2 - 4\delta < 0$ and $\tau \neq 0$ then there is a focus at the origin (show when it is a repeller);
(d) if $\delta > 0$ and $\tau = 0$ then there is a centre at the origin.

4. Find the critical points of the equation

$$\frac{d^2u}{dt^2} + u - u^2 - 2u^3 = 0$$

and determine their stability. Sketch the family of curves in the phase space.

5. Consider the Rayleigh equation

$$\frac{d^2u}{dt^2} + u = \mu\left(1 - \left(\frac{du}{dt}\right)^2\right)\frac{du}{dt}$$

for $\mu > 0$. Show that this equation admits a limit cycle. Is this equation related to Van der Pol's equation

$$\frac{d^2u}{dt^2} - \mu(1 - u^2)\frac{du}{dt} + u = 0 \quad ?$$

6. Consider the differential equation

$$\frac{d^2u}{dt^2} + f(u) = 0.$$

Show that if this equation has a critical point in the u–du/dt phase plane, where the linearised equation indicates a centre, then the nonlinear equation also has a centre at this fixed point.

7. Find and characterize the fixed points of the differential equation

$$\frac{d^2u}{dt^2} + u - \mu u^3 = 0.$$

Show that if $\mu > 0$ and

$$|u(0)| < \frac{1}{\sqrt{\mu}},$$

then the periodic solutions are possible for

$$(u(0))^2 + (u(0))^2 < \frac{\mu(u(0))^4}{2} + \frac{1}{2\mu}.$$

8. Consider the autonomous system of differential equations

$$\frac{du_1}{dt} = -u_1$$

$$\frac{du_2}{dt} = -u_2 + u_1^2$$

$$\frac{du_3}{dt} = u_3 + u_1^2.$$

Determine stable and unstable manifolds of the fixed point at $(0,0)$ and relate them to stable and unstable subspaces of the appropriate linearised system.

9. Observe the solution of the driven Duffing's equation

$$\frac{d^2u}{dt^2} + 0.1u + u^3 = B\cos t$$

for $9 \leq B \leq 10$. Change B with a stepwidth of 0.1. Try to characterise them.

10. Compute the Poincaré map of the equation of Problem 9.

11. Define a Poincaré map for the differential equations

(a) $$\frac{d^2u}{dt^2} + a\frac{du}{dt} + u + cu^3 = A\sin\frac{1}{4}t + \sin t,$$

(b) $$\frac{d^2u}{dt^2} + a\frac{du}{dt} + bu + cu^3 = A\cos\omega t\cos\Omega t,$$

where ω and Ω are incommensurate. What is a dimension of the Poincaré maps in both cases?

12. Find the power spectrum of the 'square' function:

$$f(t) = \begin{cases} a & t \in [0,\pi/2] \\ 0 & t \notin [0,\pi/2] \end{cases}.$$

Show that the average power is $f^2(t) = a^2$.

13. Compute the largest Lyapunov exponent for the equation given in Problem 9. Try to find the value of B at which chaos appears for the first time.

3. Discrete Dynamical Systems

A discrete dynamical system is a system which is discrete in time so we observe its dynamics not continuously but at given moments of time as in the case of the Poincaré map introduced in the previous chapter. The dynamics of discrete dynamical systems is usually simple enough to be explained in detail. We use these systems to describe the main phenomena of nonlinear dynamics.

3.1 Introductory Example

As an introductory example of a mechanical system which leads to a nonlinear map we have taken the ball bouncing on a vibrating table. The model which includes two parameters, serves as a conceptually easy bridge between the discrete map and important three-and four-parameter experiments like the driven nonlinear oscillator and the dynamics of the forced pendulum.

Consider the model shown in Fig. 3.1. An elastic ball is falling under the

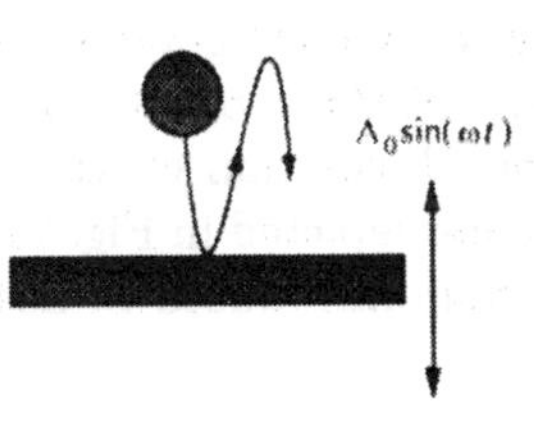

Fig. 3.1. Bouncing ball model

gravitational force onto the sinusoidally vibrating table. The equations of motion of the table and of the ball are as follows:

$$\begin{aligned} A(t) &= A_0 \sin(\omega t), \\ Z(t) &= Z_0 + Vt - \tfrac{gt^2}{2}, \end{aligned} \tag{3.1}$$

where Z_0 A_0 and ω are initial position of the ball, amplitude and frequency of the table vibration, respectively. g is the gravitational acceleration. The motion of the table $A(t)$ and of the ball $Z(t)$ are constrained through the inelastic impact [defined as $A(T) = Z(t)$]:

$$K = -\frac{(V_n(t_n) - A_n(t_n))}{(U_n(t_n) - A_n(t_n))}, \tag{3.2}$$

where U_n, V_n and A_n are, respectively, the absolute velocities of the approaching ball, the departing ball and the table. K is the coefficient of restitution and t_n is the time of the nth impact. In our further study we assume that the distance the ball move between impacts under the influence of gravity is large compared with the displacement of the table. In this case the time interval between impacts can be approximated as

$$t_{n+1} - t_n = \frac{2V_n}{g} \tag{3.3}$$

and the velocity of approach at the $(n+1)$st impact as

$$U_n(t_{n+1}) = -V_n(t_n). \tag{3.4}$$

From (3.1)–(3.4) one obtains, after nondimensionalizing, the recurrence relation between the state of the system at the $(n+1)$th and nth impacts in the form of the nonlinear discrete map,

$$\begin{aligned} \phi_{n+1} &= \phi_n + v_n \\ v_{n+1} &= K v_n - \delta \cos(\phi_n + v_n), \end{aligned} \tag{3.5}$$

where $\phi = \omega t$, $v = 2\omega V/g$ and $\delta = 2\omega^2(1+K)A_0/g$.

The analysis of the bouncing-ball example leads directly to a discrete dynamical system. Basically, the study of iterated maps has evolved from the desire to understand the behaviour of solutions of ordinary differential equations. Differential equations give rise to iterated maps in two different ways. One way is via numerical solution. Virtually any scheme for integrating a differential equation numerically (such as Runge-Kutta) reduces itself to an iterative procedure or a mapping. Another technique for reducing a differential equation to a map is appropriate when there is a surface of section or cross-section present for the flow. This occurs when all solution curves repeatedly intersect a submanifold of codimension one, as depicted in Fig. 2.11 in Sect. 2.4. The map in this case is the *Poincaré map* introduced in Sect. 2.4.

3.2 One-Dimensional Maps

Consider a map from a set S onto itself. As this map is iterated over and over again, it generates a discrete time dynamical system (or difference equation),

$$x_{n+1} = f(x_n).$$

The time takes the values $t = n = 0, 1, 2, \ldots$. The subject of dynamics is concerned with the long-time or asymptotic behaviour of the map.

Let $f(f : S \to S)$ be the map under consideration. We write f^n for f composed with itself n times. For $x \in S$, the set

$$\{x, f(x), f^2(x), \ldots\}$$

is called the orbit of x. The orbits tend to close in on certain subsets of S. If $S \in R$ this map is one-dimensional. Our goal will be to describe the dynamics of f. This means that we are interested in the behaviour of points under iteration of f. We will denote the nth iterate of f by $f^{(n)}$, i.e.

$$f^{(2)} = f \circ f, \qquad f^{(3)} = f \circ f \circ f,$$

and so forth.

The main question in dynamics is: Can one predict the fate of all orbits of f? That is, what can be said about the behaviour of $f^{(n)}(p)$ as $n \to \infty$? As we shall see, even for simple quadratic maps of the real line the answer to this question is quite difficult, but extremely interesting.

Example. Consider the following simple example from ecology. Suppose the population of a single species reproducing in a controlled environment has population at generation n given by P_n. To keep the numbers manageable, let us assume that P_n represent the percentage of some *a priori* upper bound for the population, so $0 \leq P_n \leq 1$. For the ecologist, the important problem is to construct a mathematical model which allows him or her to predict the ultimate behaviour of the population. Will the population die out or will it tend to stabilize at some limiting value? Or will it change cyclically or behave in some other more random fashion?

One of the simplest models of population growth used in ecology is the *logistic equation.* This equation is given by

$$P_{n+1} = k(P_n(1 - P_n).$$

Here k is a constant which depends on ecological conditions. Given this equation, it would seem straightforward to predict the ultimate behaviour of P_n given some initial population P_0. As we shall see, however, this is far from the case for certain k-values. Note that if we let $f(x) = kx(1 - x)$, then the solution of the logistic equation for a given initial population P_0 is equivalent to computing the orbit of P_0:

$$P_0, \quad P_1 = f(P_0), \quad P_2 = f^{(2)}(P_0), \ldots \quad . \qquad \square$$

Example. Now let us consider the dynamics of $f : \mathcal{R} \to \mathcal{R}$. Computing orbits is most easily done with a computer, or, in our simple setting, even with a calculator. For example, if $f(x) = x^2$, it is clear that

$$
\begin{aligned}
f^{(n)}(x) &\to \infty && \text{if } |x| > 1, \\
f^{(n)}(x) &\to 0 && \text{if } |x| < 1, \\
f^{(n)}(1) &= 1 && \text{for all } n, \\
f^{(n)}(-1) &= 1 && \text{if } n \geq 1,
\end{aligned}
$$

that is, all orbits tend asymptotically either to ∞ or to 0, except $x^* = 1$, which is a *fixed point*, or -1, which is eventually fixed. The point $x^* = 0$ is also a fixed point. There is a difference between the fixed points at 0 and 1. 0 is an *attracting fixed point* or *sink*, since nearby points have orbits which tend to 0. The point 1, on the other hand, is a *repelling fixed point* or *source*, since the nearby orbits tend away from 1. □

Example. Another simple example is provided by the square-root function $T(x) = \sqrt{x}$. Since

$$T^n(x) \to 1$$

for all $x > 0$, 1 is an attracting fixed point. □

Example. $S(x) = \sin x$ in radians similarly has 0 as an attracting fixed point which attracts all orbits. Of course, $\pi, 2\pi, 3\pi, \ldots$ are all eventually fixed points. Note that orbits take a very long time to reach 0 under iteration of S; although 0 is an attracting fixed point, it is a "weak" attractor for reasons we will explore in a moment. □

Example. A final example which is quite interesting for its unexpected result is the cosine. What happens when $C(x) = \cos x$ is iterated? With a calculator one sees immediately that all orbits tend to 0.73908... . This comes as a surprise to the uninitiated, but we will see below that this point is nothing but another attracting fixed point. □

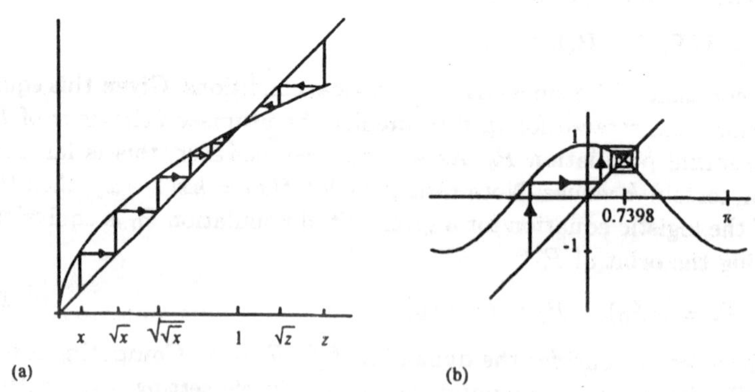

Fig. 3.2. Graphical analysis of functions. **(a)** $T(x) = \sqrt{x}$. **(b)** $C(x) = \cos x$

A simple technique for studying one-dimensional dynamics is *graphical analysis.* This technique allows us to predict the qualitative behaviour of orbits from the knowledge of the graph of a function. To compute the orbit of a point x under iteration of f, one simply draws the diagonal $y = x$ as well as the graph of f, and then repeats the following procedure over and over. If we draw a vertical line from (x, x) to the graph of f, followed by a horizontal line back to the diagonal, then we reach the point $[f(x), f(x)]$. Repeating this process yields the point $[f^2(x), f^2(x)]$. Continuing, we see that the orbit of x is displayed along the diagonal. Fig. 3.2 displays graphical analysis as applied to $T(x) = \sqrt{x}$ and $C(x) = \cos x$.

Graphical analysis allows us to understand what makes certain fixed points attracting and others repelling. If the derivative of f at the fixed point is less than 1 in absolute value, then the fixed point is attracting. If the absolute value of the derivative is larger than 1, then the fixed point is repelling. See Fig. 3.3 for details.

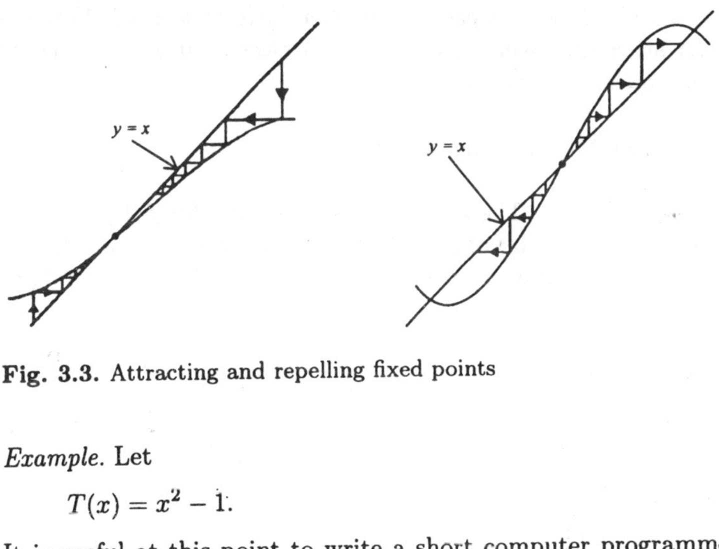

Fig. 3.3. Attracting and repelling fixed points

Example. Let

$$T(x) = x^2 - 1.$$

It is useful at this point to write a short computer programme to iterate a function like T. The code is quite easy to write (usually a few lines long) and immediately produces a list of the orbit of an initial input x_0. One sees immediately that certain orbits tend to ∞, while other orbits tend to oscillate between values near 0 and -1. Graphical analysis (Fig. 3.4) confirms this. T has two fixed points, the roots of

$$x^2 - 1 = x.$$

Let x_+ denote the positive fixed point. Then x_+ is repelling and, indeed, if $|x| > x_+$, the orbit of x tends to ∞. On the other hand, we have

$$T(0) = -1, \qquad T(-1) = 0$$

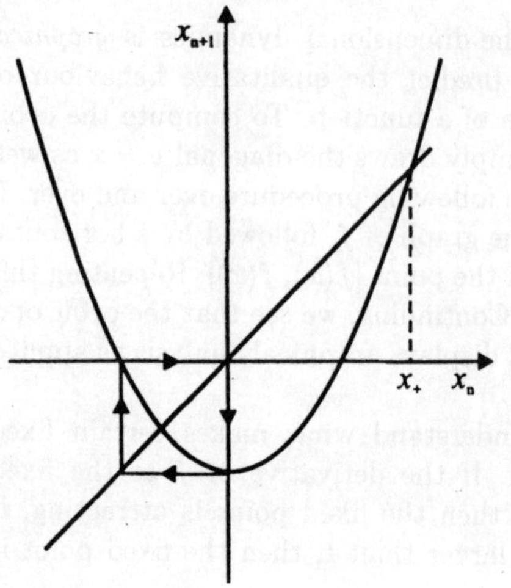

Fig. 3.4. Graphical analysis of $T(x) = x^2 - 1$

so we say that 0 and -1 lie on a *periodic orbit* or *cycle of period* 2. Note that this orbit is an attracting orbit, because if x is close to 0 or -1, then the orbit of x tends to the periodic orbit. □

Now we introduce the following definitions:

Definition. Let $f : \mathcal{R} \to \mathcal{R}$. The point x_0 is a fixed point for f if $f(x_0) = x_0$. The point x_0 is a periodic point of period n for f if $f^n(x_0) = x_0$ but $f^i(x_0) \neq x_0$ for $0 < i < n$. The point x_0 is eventually periodic if $f^n(x_0) = f^{n+m}(x_0)$, but x_0 is not itself periodic.

Definition. A periodic point x_0 of period n is attracting if $|(f^n)'(x_0)| < 1$. The periodic point x_0 is repelling if $|(f^n)'(x_0)| > 1$. x_0 is neutral if $|(f^n)'(x_0)| = 1$.

Let us consider with more details the dynamical behaviour exhibited by the quadratic logistic map,

$$x_{n+1} = ax_n(1 - x_n). \tag{3.6}$$

If $3 > a > 1$, the fixed point at $x^* = 1 - 1/a$ is an attractor as shown in Fig. 3.5a, and the system settles down to the stable point made familiar by countless discussions in elementary mathematics courses. At $a = 3$ the system bifurcates, to give a cycle of period 2 (Fig. 3.5b), which is stable for $1+\sqrt{6} > a > 3$. As a increases beyond this, successive bifurcations give rise to a cascade of period doublings, producing cycles of periods 4 (Fig. 3.5c), and then 8, 16, ..., 2^n. With further increase of a we observe a chaotic regime, in which trajectories look like the sample functions of random processes (Fig. 3.5d).

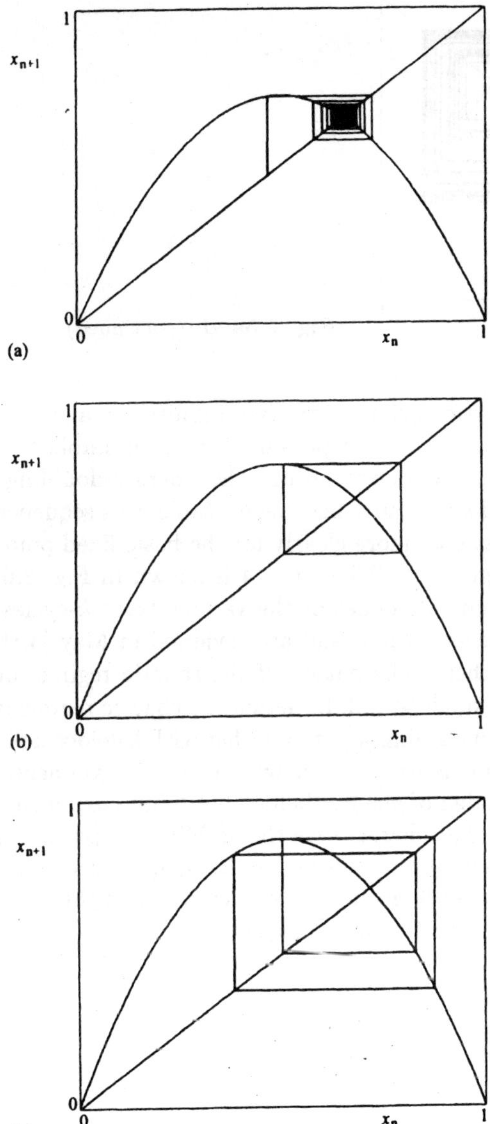

Fig. 3.5a–d. Attractors of the logistic map. **(a)** Fixed point. **(b)** Period 2 orbit. **(c)** Period 4 orbit. **(d)** Chaotic trajectory

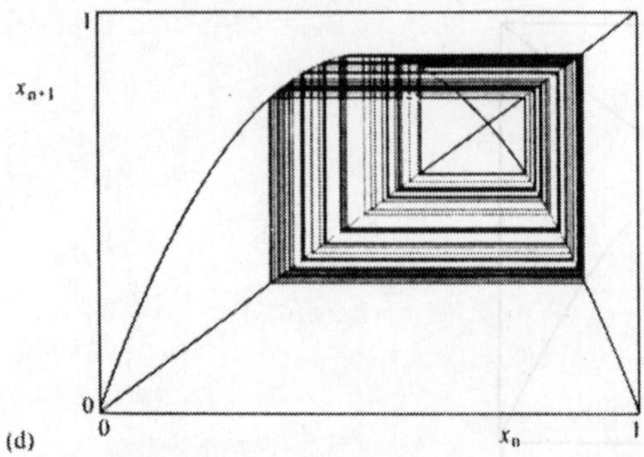

Fig. 3.5a–d. (continued)

In detail, the apparently chaotic regime comprises infinitely many tiny windows of a-values, in which basic cycles of period k are born stable (accompanied by unstable twins), cascade down through their period-doublings to give stable harmonics of period $k \times 2^n$, and become unstable; this sequence of events recapitulates the process seen more clearly for the basic fixed point of period 1. The bifurcation diagram for $2.9 < a < 4$ is shown in Fig. 3.6. The details of these processes, and catalogues of the various basic k-cycles, have been given independently several times and are reviewed in May [3.1], Collet and Eckmann [3.2] and others. The nature of the chaotic regime for such 'maps of interval' is often misunderstood. In detail, the chaotic regime is largely a mosaic of stable cycles, one giving way to another with kaleidoscopic rapidity as a increases. This point is exemplified by 'Lyapunov exponents' that are often computed as an index of chaotic behaviour. These exponents are analogous to the eigenvalues that characterise the stability properties of simpler systems. They are typically calculated by iterating difference equations, such as (3.6), and calculating the geometric average value of the slope of the map at each iterate: Hence, for the difference equation

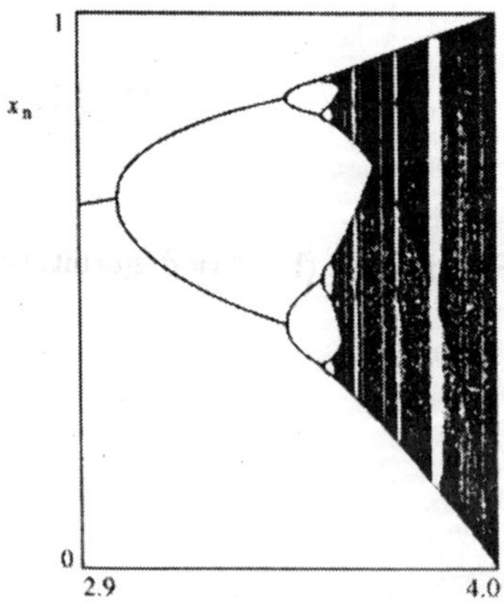

Fig. 3.6. Bifurcation diagram of the logistic map

$$x_{n+1} = f(x_n)$$

the Lyapunov exponent λ is given by

$$\lambda = \lim_{n\to\infty} \left\{ \frac{1}{n} \sum_{i=0}^{n-1} \ln \frac{\mathrm{d}f(x_n)}{\mathrm{d}x} \right\}.$$

For generically quadratic maps, there are unique attractors for most values of a in the chaotic regime. Therefore, this calculation, if carried out exactly, or if the iterations are carried on long enough, shows that λ is typically positive in the chaotic regime. The plot of Lyapunov exponent for logistic map is shown in Fig. 3.7. In the bifurcation diagram we can distinguish between the bifurcation domain for $1 < a < a_\infty$, where Lyapunov exponent is negative (excluding period-doubling bifurcation points $a_1, a_2, \ldots$ where it is zero) and the chaotic domain for $a_\infty < a \leq 4$ where λ is mostly positive. In the 'chaotic domain' we can also observe a-windows where λ is negative and the sequence of iterates is periodic.

Period-doubling bifurcation leading to chaos has an interesting property. Consider the following ratio:

$$\rho_n = \frac{a_n - a_{n-1}}{a_{n+1} - a_n}.$$

It was observed by Feigenbaum [3.3] that

$$\lim_{n\to\infty} \rho_n = \delta,$$

where $\delta = 4.66920\ldots$ Feigenbaum's numerical computations and analytical results of Collet, Eckmann and Lanford [3.2] show that δ has some universal properties.

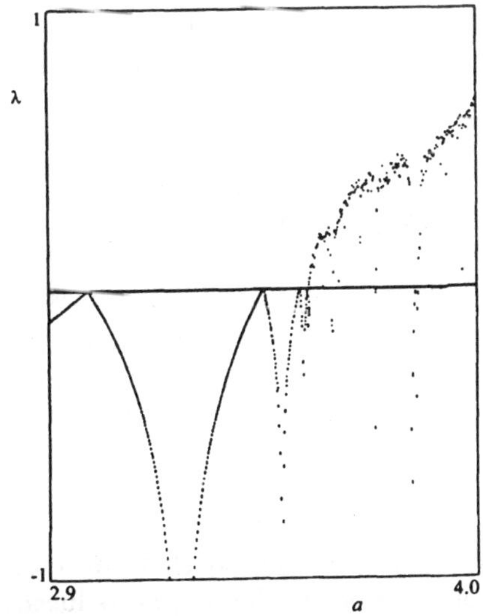

Fig. 3.7. Lyapunov exponent of the logistic map

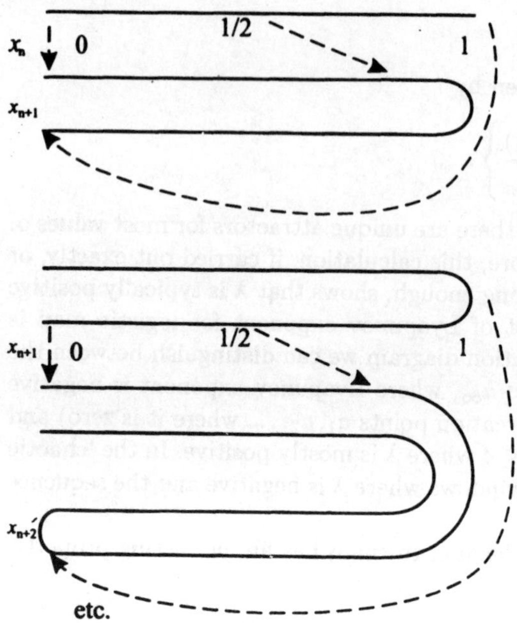

Fig. 3.8. The stretching and folding mechanism for the logistic map

Finally, we note that chaotic behaviour of a logistic map can be explained using the same stretching and folding mechanism as that mentioned in Chap. 2. As we already know this mechanism is necessary to keep chaotic trajectories within a finite volume of phase space, despite the exponential divergence of neighbouring states. The logistic map stretching (divergence of nearby trajectories) and folding (confinement to bounded space) are shown in Fig. 3.8. For $a = 4$ the logistic map has a maximum value of 1 for $x_n = 1/2$ and the values $x_n \in (1/2, 1)$ as well as values $x_n \in (1/2, 1)$ map to $x_{n+1} \in (0, 1)$. Therefore both intervals of x_n are stretched by a factor 2, but because the order of mapping is opposite, the second stretched interval is folded onto the first stretched interval.

3.3 Bifurcations of One-Dimensional Maps

In the previous section we observed that with the increase of a the logistic map changes its behaviour qualitatively. Such a qualitative change in the dynamics which occurs as a system parameter varies is called a bifurcation. In this section we describe the bifurcations characteristic of one-dimensional maps,

$$x_{n+1} = f(x_n, a),$$

where f is continuous and differentiable. Without generality we shall consider bifurcations of fixed points. Bifurcations of a period n orbit can be reduced

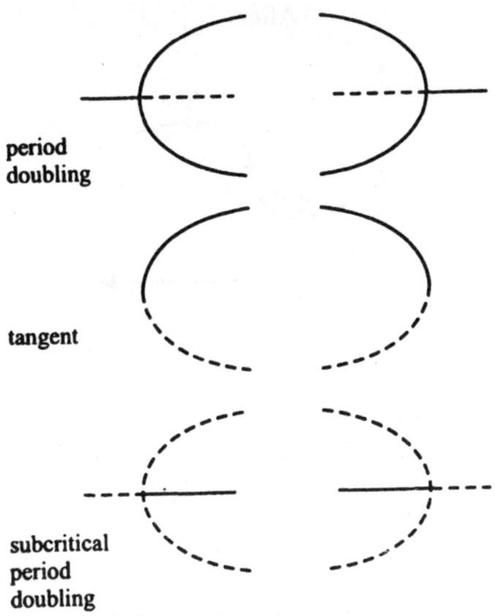

Fig. 3.9. Generic bifurcations of one-dimensional map

to consideration of the n times iterated map $f^{(n)}$ for which each point of the period n orbit is a fixed point.

We shall consider only generic bifurcations, i.e., bifurcations, whose basic character cannot be altered by arbitrarily small continuous and differentiable perturbations.

There are three generic types of bifurcations of continuous and differentiable one-dimensional maps,

- the period doubling bifurcation,
- the tangent bifurcation,
- the subcritical period-doubling bifurcation.

These are illustrated in Fig. 3.9 in the way they are visible in bifurcation diagrams. Dashed lines are used for unstable orbits and solid lines for stable orbits. The parameter a is assumed to increase to the right. Additionally in Fig. 3.9 we have defined forward and backward bifurcations.

In Fig. 3.10 we showed how three forward bifurcations can occur as the shape of the map changes with increasing control parameter a. Note that at the bifurcation point $x_0, |f'(x_0)| = 0$, thus x_0 is neutral.

3.4 One-Dimensional Maps and Higher-Dimensional Systems

In this section we show how the methods developed for one-dimensional maps can be useful for investigation of chaotic behaviour of higher-dimensional systems.

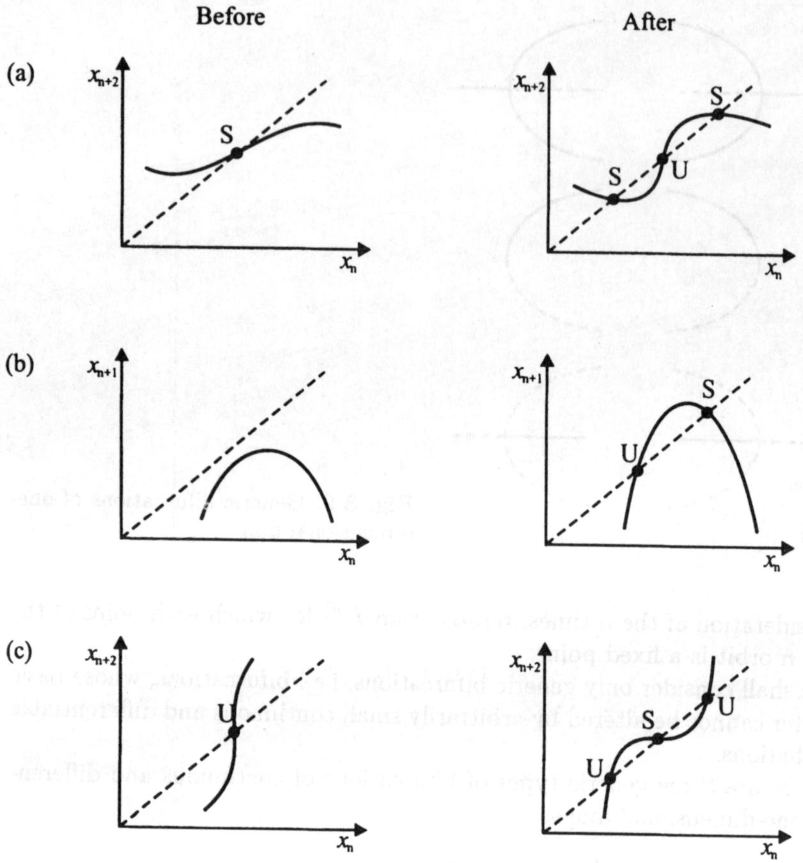

Fig. 3.10. Behaviour of the map (3.6) before and after bifurcation (a) to (c). (a) period-doubling (b) tangent (c) subcritical period-doubling

In Sect. 2.4 we introduced the idea of Poincaré maps, which allow one to replace the consideration of an N-dimensional continuous dynamical system by an equivalent $(N-1)$-dimensional system. As chaotic behaviour can be observed in at least three-dimensional systems, the equivalent Poincaré maps are at least two-dimensional. However, in a number of dynamical systems we can define one-dimensional maps similar to that considered in the previous section. These maps can significantly simplify the investigations of chaotic behaviour of the considered dynamical system.

As an example let us consider a Lorenz equation

$$\frac{\mathrm{d}x}{\mathrm{d}t} = -\delta(x-y)\,,$$
$$\frac{\mathrm{d}y}{\mathrm{d}t} = -xz + rx - y\,, \tag{3.7}$$

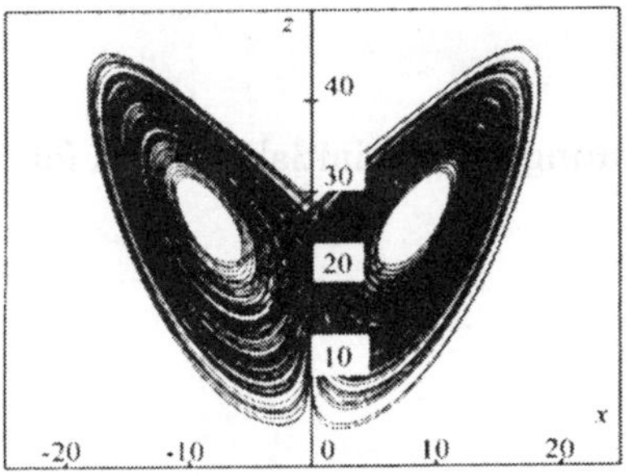

Fig. 3.11. Phase space trajectory of Lorenz system (3.7)

$$\frac{dz}{dt} = xy - bz ,$$

where δ, r and b are dimensionless parameters [3.4]. It is well-known that for $\delta = 10, b = 8/3$ and $r = 28$, equation (3.7) shows chaotic behaviour. Fig. 3.11 shows a projection of the phase-space trajectory onto the yz plane. The points C_1 and C_2 represent projections of fixed points which are unstable for the considered parameter values. We can observe that the solution spirals outward from one of the fixed points C_1 or C_2 for some time, then switches to spiraling outward from the other fixed point. This pattern repeats forever with the number of revolutions around a fixed point before switching appearing to vary in a random manner.

Let m_n be the nth maximum of the function $z(t)$. If we plot m_{n+1} versus m_n we obtain the one-dimensional map shown in Fig. 3.12, whose dynamics is very similar to the so-called tent map

$$x_{t+1} = 1 - 2\left|\frac{1}{2} - x_t\right| .$$

Another example of a one-dimensional map which describes chaotic behaviour of a higher-dimensional system will be given in Sect. 5.4.

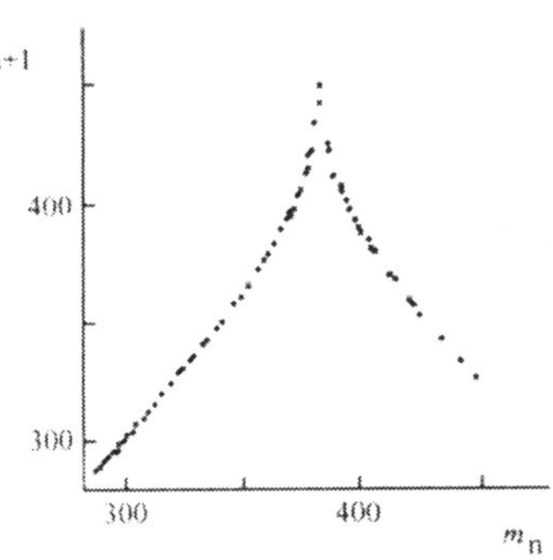

Fig. 3.12. m_{n+1} versus m_n map

Problems

1. Observe different types of orbits by iterating various initial x-values for

$$T_c(x) = x^2 + c$$

   where
   (a) $c = -2.1$,
   (b) $c = -1.3$ (period 4),
   (c) $c = -1.38$ (period 8),
   (d) $c = -1.395$ (period 16),
   (e) $c = -1.755$ (period 3),
   (f) $c = -2$,
   (g) $c = -3$.
2. Compute an appropriate bifurcation diagram for the logistic map (3.6). Define the distance d_n of the points in 2^n-periodic orbit which are the closest to $x = 0.5$ and show that for large n the ratio $d_n/d_{n+1} = 2.5029\ldots$
3. Observe the structure of windows in the chaotic domain. Notice that a windows are characterized by periodic p orbits ($p = 3, 4, 6, \ldots$) with successive bifurcations p, $p2^1$,$p2^3$ etc. Try to show that corresponding a values fulfill Feigenbaum's relation.
4. Compute the bifurcation diagram of a map $x_{n+1} = x_n^2 + c$, $-2 < c < 0.3$.
5. Expand the scale of c for the bifurcation diagram of the previous problem in the region of period-doubling bifurcation in order to observe as many period-doubling bifurcations as possible. Try to verify approximately the Feigenbaum number.
6. Expand the scale of c for the bifurcation diagram of Problem 3 for $-1.8 < c < -1.7$. Notice a period 3 window and examples of period-doubling bifurcations. Do you notice any similarities between these bifurcations and the period-doubling bifurcations observed in Problem 4.

4. Fractals

Fractals, objects with noninteger dimension, may at first sight seem to be unlikely candidates for any practical applications. In this chapter we introduce basic examples and properties of fractal sets starting with a classical example of the Cantor set and introduce different definitions of its dimension. Later we discuss the application of the fractal concept to dynamics and show that it is very useful in the description of strange chaotic attractors.

4.1 The Cantor Set

We describe the Cantor set by a construction using tremas (Edgar [4.1]). The *triadic Cantor dust* is a subset of the line $\mathcal{R}$. A sequence of approximations is first defined. Start with the closed interval $C_0 = [0, 1]$. Then the set C_1 is obtained by removing the 'middle third' from $[0, 1]$, leaving $[0, 1/3] \cup [2/3, 1]$. The next set C_2 is defined by removing the middle third of each of the two intervals of C_1. This leaves

$$C_2 = [0, 1/9] \cup [2/9, 1/3] \cup [2/3, 7/9] \cup [8/9, 1],$$

and so on (see Fig. 4.1). The *triadic Cantor dust* is the 'limit' C of the sequence C_n of sets. The sets decrease: $C_0 \subseteq C_1 \subseteq C_2 \subseteq \cdots$. So we will define the limit to be the intersection of the sets,

$$C = \bigcap_{k \in \mathcal{N}} C_k.$$

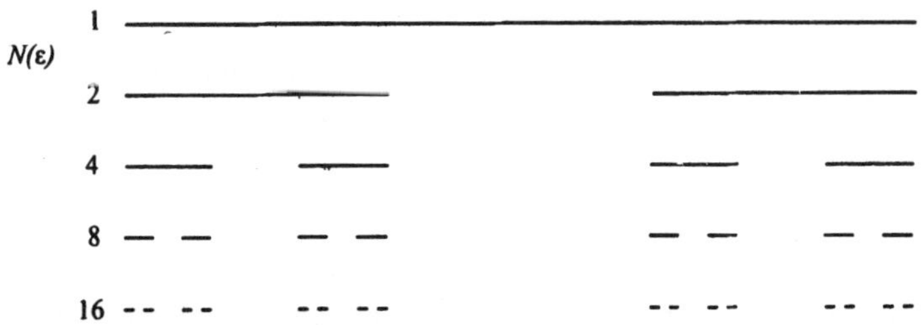

Fig. 4.1. The triadic Cantor dust

The parts that are removed are called *tremas*.

The sequence of sets is defined recursively. This means that it will often be easy to prove facts about the sets by induction. For example, the set C_k consists of 2^k disjoint closed intervals, each of length $(1/3)^k$. So the total length of C_k, the sum of the lengths, is $(2/3)^k$. The limit is

$$\lim_{k\to\infty}\left(\frac{2}{3}\right)^k = 0.$$

So the total length of the Cantor dust itself is zero. (The mathematical version of total length is called Lebesgue measure).

Definition. Let $S \subset \mathcal{R}$ be a bounded nonempty open set. It can be represented as the sum of a finite or a countably infinite number of disjoint open intervals whose end points do not belong to S, i.e.

$$S = \sum_k (a_k, b_k).$$

The Lebesque measure of the open set S is

$$\mu(S) = \sum_k (b_k - a_k).$$

Therefore the total length is not a very useful way to compute the size of C. We will see that this is related to the fact that the fractal dimension of C is smaller than 1.

Let us consider more carefully which points constitute the Cantor set. If $[a, b]$ is one of the closed intervals that make up one of the approximations C_k, then the endpoints a and b belong to *all* of the future sets $C_m, m \geq k$ and therefore belong to the intersection C (this can be proved by induction). Taking all the end points of all the intervals of all the approximations C_k, we get an infinite set of points, all belonging to C (this set of endpoints is however only a countable set). But it is important to note that there are points in C other than these end points.

Example. The point $1/4$ is not an endpoint of any interval of any set C_k. But the point $1/4$ belongs to C. □

It is also important to note that C is not like the usual sets of elementary geometry. At first, it is likely to tax your powers of geometrical visualization. Here are a few tips to help:

(1) The set C contains no interval (of positive length).
(2) The set C has no isolated points: that is, if $a \in C$, then for every $\epsilon > 0$, no matter how small, the interval $(a - \epsilon, a + \epsilon)$ contains points of C in addition to a.

(3) The set C is closed: that is, if $a \in \mathcal{R}$ has the property that every interval of the form $(a-\epsilon, a+\epsilon)$ intersects C, then $a \in C$.

There is a convenient way to characterise the elements of the triadic Cantor dust in terms of the expansions in base 3.

First we will review the standard facts concerning base 3. You know how expansions in the usual base 10 work. Base 3 is of course similar. Every positive integer x has a unique representation

$$x = \sum_{j=0}^{M} a_j 3^j,$$

where the 'digits' a_j are chosen from the list 0, 1, 2. For example

$$15 = 0 \cdot 3^0 + 2 \cdot 3^1 + 1 \cdot 3^2.$$

We will sometimes write simply $15 = (120)_3$. It is understood that the subscript specifying the base is always written in base 10.

Similarly, we may represent fractions: Every number x between 0 and 1 has a representation in the form

$$x = \sum_{j=-\infty}^{-1} a_j 3^j$$

with digits 0, 1, 2. These are written with a 'radix point' (or 'ternary point' in this case):

$$7/9 = (0.21)_3$$

$$1/4 = (0.02020202\cdots)_3 \qquad \textit{(repeating)}$$

$$\sqrt{2} = (1.1020112\cdots)_3 \qquad \textit{(nonrepeating)}.$$

Some numbers (rational numbers of the form $a/3^k$) admit two different expansions for example,

$$1/3 = (0.1000000\cdots)_3 = (0.0222222\cdots)_3.$$

Proposition. Let $x \in [0,1]$. Then x belongs to the triadic Cantor dust C if and only if x has a base 3 expansion using only the digits 0 and 2.

Proof. The first place to the right of the ternary point is 1 if and only if x is between

$$(0.1000000\cdots)_3 = 1/3, \quad \text{and} \quad (0.1222222\cdots)_3 = 2/3.$$

The first trema is the interval $(1/3, 2/3)$. After this trema is removed, we have C_1. (The numbers 1/3 and 2/3 each have two expansions, one with 1 in the first place, and one without. So they should not be removed.) Therefore C_1

contains exactly the numbers in $[0,1]$ that have a base 4 expansion not using 1 in the first place. The second place of a number x in C_1 is 1 if and only if x belongs to one of the second-level tremas $(1/9, 2/9)$ or $(7/9, 8/9)$. When these tremas are removed we have C_2. So C_2 contains exactly the numbers in $[0,1]$ that have base 3 expansion not using 1 in the first or second place. Continuing in this way, we see that the points remaining in $C = \bigcap_{k \in \mathcal{N}} C_k$ are exactly the numbers in $[0,1]$ that have base 3 expansion not using 1 at all.

The Cantor dust is uncountable. This follows from the representation that proved, together with the observation, that each real number has at most two representations base 3. Actually, for numbers in the Cantor dust, two different sequences of 0s and 2s always represent real numbers.

4.2 Fractal Dimensions

Now let us introduce what are known as *fractal dimensions*. First we give the definition of the two most commonly used fractal dimensions: the capacity and the Hausdorff dimension.

The usual notion of dimension of a set corresponds to the number of parameters one has to use to indicate the position of a point in this set. The so called *topological dimension* d_{T} takes only positive integer values.

Hausdorff in [4.2] suggested another definition of the dimension based on the generalization of the notion of length.

Let S be a set in a metric space $\mathcal{R}^n$. We then define the Hausdorff measure of s, indexed by the parameter $\delta \in \mathcal{R}$, in the following way:

$$l_\delta(S) = \lim_{\epsilon \to 0+} l_{\delta,\epsilon}(S),$$

where

$$l_{\delta,\epsilon}(S) = \inf_{K(\epsilon)} \sum_{B_i \in K(\epsilon)} |B_i|^\delta,$$

and where the lower bound is taken over all the coverings $K(\epsilon)$ of the set S made of balls $(B_i)_i$ of diameter smaller then ϵ. The *Hausdorff dimension* of S, $d_{\mathrm{H}}(S)$, is then defined as the unique value of δ such that $l_\delta(S)$ is finite

$$\delta > d_{\mathrm{H}}(S) \to l_\delta(S) = 0,$$

$$\delta < d_{\mathrm{H}}(S) \to l_\delta(S) = +\infty.$$

Let us note that d_{H} can take noninteger values. The Hausdorff measure associated to the dimension d_{H} is $l_{d_{\mathrm{H}}}$. It is a generalization of the Lebesque measure in $\mathcal{R}^n$. Thus, in order to evaluate the relative 'size' of two given sets, one just needs to compare their Hausdorff dimensions and, if they are equal, the values of their Hausdorff measures. Although the Hausdorff dimension is very well defined mathematically, it is generally hard to estimate numerically. To circumvent this difficulty, a more practical definition of the dimension of a set is generally used.

The *capacity* or *box-counting dimension* has been introduced by Kolmogorov [4.7]. Let S be a subset of $\mathcal{R}^n$ and $K(\epsilon)$ a covering of S with balls of size ϵ. Let $N(\epsilon)$ be the number of balls in $K(\epsilon)$. The capacity of S, $d_c(S)$, is then defined as the limit

$$d_c(S) = \lim_{\epsilon\to 0+} \sup \frac{\ln N(\epsilon)}{\ln(1/\epsilon)}.$$

In order to understand three introduced dimensions and differences between them we consider the following examples.

Example. The Hausdorff dimension and the capacity dimension of the empty set are equal:

$$d_H = d_c = -\infty,$$

and in this case the topological dimension d_T is not defined. □

Example. For sets such as a point, a segment, a surface, the Hausdorff dimension d_H and the capacity dimension d_c are equal to the topological dimension which is respectively 0, 1 and 2. □

Example. Consider again the triadic Cantor set. At the step n of the construction process, the set is made of 2^n intervals of equal length 3^{-n}. Thus , the Lebesque measure of the triadic Cantor set obtained when $n \to \infty$ is 0, and its topological dimension $d_T = 0$. Let us consider the covering corresponding to the step n of the construction process. As we already pointed out, it is made of

$$N(\epsilon) = 2^n$$

intervals of size $\epsilon = 3^{-n}$. When ϵ goes to 0 ($n \to \infty$), one thus deduces that the capacity of the triadic Cantor set is

$$d_c = \frac{\ln 2}{\ln 3}.$$

The Hausdorff dimension can be obtained in a similar way and is found to be the same as d_c, so the triadic Cantor set is then characterized by the following dimensions

$$d_T = 0,$$

$$d_H = d_c = \frac{\ln 2}{\ln 3}.$$ □

Generally one can show that

$$d_{\mathrm{T}} \leq d_{\mathrm{H}} \leq d_{\mathrm{c}}.$$

A classical example for which $d_{\mathrm{H}} < d_{\mathrm{c}}$ is the set of all the rational numbers in $[0,1]$. As it is dense in $[0,1]$, the capacity is $d_{\mathrm{c}} = 1$. However the Hausdorff dimension is equal to the topological dimension which is 0.

Capacity dimension gives the scaling of the number of cubes needed to cover the attractor. In the case of the strange attractor the frequency with which different cubes are visited is vastly different from cube to cube, sc ͡or very small ϵ, it is common that only a very small number of the cubes need to cover the chaotic attractor containing the vast majority of the natural measures on the attractor. Typical orbits spend most of their time in a small minority of those cubes that are needed to cover the attractor. To take into account different natural measures of the cubes it is necessary to introduce another definition of dimension, the *information dimension.*

Definition. The quantity d_{I} is given by

$$d_{\mathrm{I}} = \lim_{\epsilon \to 0} \frac{\sum_{i=1}^{N(\epsilon)} \mu_i \ln \mu_i}{\ln(1/\epsilon)},$$

where $\mu_i = 1/N(\epsilon)$ is called the information dimension. μ_i can be considered as a probability with which trajectory visits i-cube.

Another useful dimension is the *correlation dimension.*

Definition. Let z_k be a trajectory on the attractor. Compute the correlation integral

$$C(\epsilon) = \lim_{k \to \infty} \frac{1}{K^2} \sum_{i,j}^{K} U(\epsilon - |z_i - z_j|),$$

where $U(.)$ is the unit step function. The quantity $C(\epsilon)$ may be shown to scale with ϵ in the following way

$$d_{\mathrm{corr}} = \lim_{\epsilon \to 0} \frac{\ln(\epsilon)}{\ln \epsilon},$$

where the quantity d_{corr} is called the correlation dimension [4.8].

The correlation dimension is very easy to estimat from experimental data usually based on a single time series (see Sect. 6.6).

The dimension of the fractal chaotic attractor can also be associated with Lyapunov exponents. Let the typical trajectory on an attractor be characterised by the following Lyapunov exponents λ_1, λ_2, ..., λ_n, ($\lambda_i \leq \lambda_{i+1}$, $i = 1, 2, ..., n-1$). The *Lyapunov dimension* d_{L} is defined in the following way.

Definition. The quantity d_L is given by

$$d_L = j + \frac{\sum_{i=1}^{j} \lambda_i}{|\lambda_{j+1}|},$$

where j is the maximum index for which

$$\sum_{i=1}^{j} \lambda_i \le 0.$$

Kaplan and Yorke [4.5] showed that for typical attractors the Lyapunov dimension is equal to the information dimension (the so-called *Kaplan–Yorke conjecture*).

4.3 Fractal Sets

Originally Mandelbrot [4.6] suggested that a fractal set should be defined as a set whose Hausdorff dimension is strictly greater than its topological dimension. This definition is adequate for a lot of sets, but there exists a whole class of sets (of the same type as the tradic Cantor set) whose Lebesque measure is finite and whose Hausdorff dimension $d_H = 1$ is equal to their topological dimension.

Example [4.3]. Consider the rational numbers in the interval $[0,1]$. These numbers are dense in $[0,1]$, since any irrational number can be approximated by a rational number to arbitrary accuracy. The rationals are also countable, since we can arrange them in a linear ordering such as

$$\frac{1}{2}, \frac{1}{3}, \frac{2}{3}, \frac{1}{4}, \frac{3}{4}, \frac{1}{5}, \frac{2}{5}, \frac{3}{5}, \frac{4}{5}, \frac{1}{6}, \frac{1}{7}, \dots.$$

To the nth rational on this list (denoted by s_n), we now associate an interval

$$I_n = \left(s_n - \frac{\eta}{2}\left(\frac{1}{2}\right)^n, s_n + \frac{\eta}{2}\left(\frac{1}{2}\right)^n \right)$$

of length $2^{-n}\,\eta$. We are interested in the set S formed by taking the interval $[0,1]$ and then successively removing $I_1, I_2, \dots, I_n$ in the limit $n \to \infty$.

Since the total length of all the removed interval set is

$$\sum_{n=1}^{\infty} \left(\frac{1}{2}\right)^n \eta = \eta$$

we have the Lebesque measure of S, denoted as $\mu(S)$ fulfilling

$$\mu(S) > 1 - \eta,$$

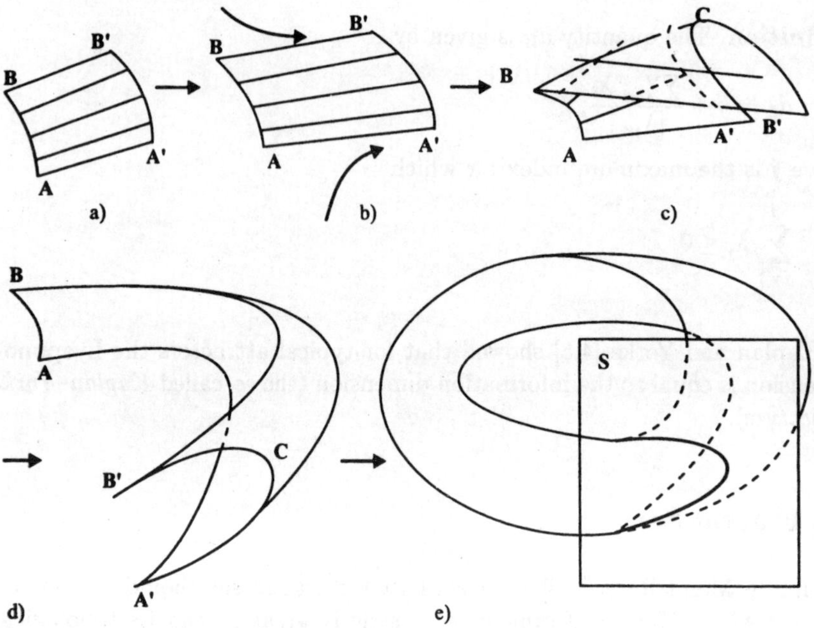

Fig. 4.2a–e. Geometrical structure of a chaotic attractor

which is positive if $\mu < 1$. The greater-than symbol appears because some of the removed intervals overlap. As the Lebesque measure of S is positive, its capacity dimension d_c is the same as the space in which S lies, i.e., $d_c = 1$ and equal to the Hausdorff d_H and topological d_T dimensions. □

Sets like the ones introduced in the above example are called *fat fractals* and can be obtained as follows: The set S lying in an n-dimensional Euclidean space is a fat fractal if, for every point $x \in S$, and every $\epsilon > 0$, a ball of radius ϵ centred at the point x contains a nonzero volume of points in the set and a nonzero volume outside the set.

As fat fractals are not fractals according to the definition given at the beginning of this section, we will rather say that a fractal set is a set which has some self-similar properties in the sense that its singular structure is the same at any scale. This definition is much wider than the one based on the Hausdorff dimension and, as we will see, can be easily applied to strange attractors.

To show that strange attractors have fractal structure let us consider a 3-dimensional chaotic attractor A on which a typical trajectory is characterised by the following Lyapunov exponents: $\lambda_1 > 0$, $\lambda_2 = 0$, $\lambda_3 < 0$ and

$$\sum_{i=1}^{3} \lambda_i < 0.$$

We recall the stretching and folding model of the chaotic attractor introduced in Chap. 2 and consider it in more detail. Consider the interval AB of which almost all points belong to the attractor A and analyse its evolution in time (schematically shown in Fig. 4.2). For small t we observe practically no changes in geometrical structure; points A and B are transposed respectively into A' and B' (Fig. 4.2b). As the sum of all Lyapunov exponents is negative, the length of $A'B'$ cannot increase to infinity, hence we observe folding of the surface $ABA'B'$ (Fig. 4.2c,d) and creation of the arch $A'CB'$. The geometrical structure of an attractor can be achieved by connecting points A' and B' (by "pressing") and "sticking" point $A' = B'$ with A and C with B. As a result we have the structure shown in Fig. 4.2e. Due to the continuity of trajectories on the strange attractor it is necessary for an infinite number of surfaces like the one in Fig. 4.2e to exist in the real attractor. To detect them consider the Poincaré map obtained by the cross-section S shown in Fig. 4.3. By enlarging parts S' and S'' we can detect the self-similar structure of the attractor showing that the attractor is a fat fractal set.

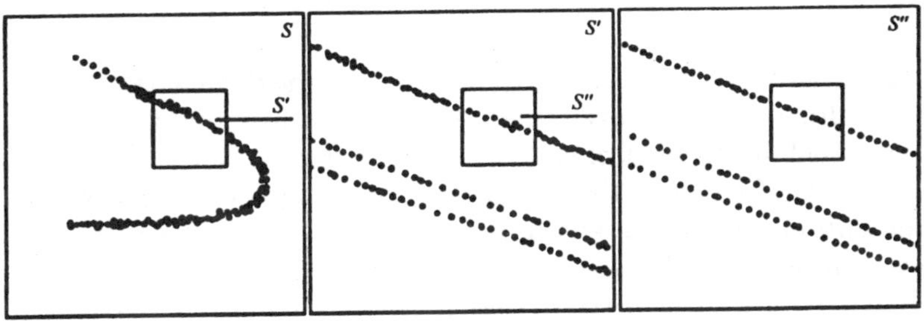

Fig. 4.3. Poincaré map of the attractor of Fig. 4.2

4.4 Smale Horseshoe

Smale [4.9] introduced the horseshoe map as a motivating example for understanding a large class of dynamical systems. This map is a two-dimensional mapping that illustrates the stretching and folding of the finite sized space as a primary mechanism for allowing sensitivity to initial conditions.

The horseshoe map is specified geometrically in Fig. 4.4. The map takes the square S, uniformly stretches it vertically by a factor $\mu > 2$ and uniformly compresses it horizontally by a factor $\mu \in (0, 1/2)$. Then the long strip is bent into a horseshoe shape with all the bending deformations taking place in the uncrossed-hatched regions of Fig. 4.4. Then the horseshoe (transformed set $f(S)$) is placed on the top of the original square, so that the map is now confined to a subset of the original unit square. If the entire sequence of

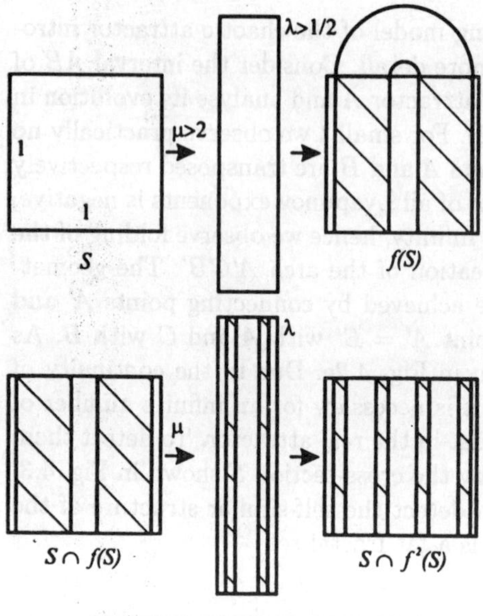

Fig. 4.4. Construction of the horseshoe map (two iterations)

operations is repeated, then four stripes appear from the original two, and so on. The repetition of the process n times leads to 2^n stripes, and a cut across the stripes would, in the limit of large n, lead to a fractal. Note that in each iteration a certain fraction of the original area of the square S is mapped to the region outside the square, and in the ˊmit $n \to \infty$ almost every initial condition with respect to Lebesque measure leaves the square.

A configuration similar to the horseshoe map occurs in the phase space of dynamical systems where there are regions with strong contraction and expansion. In the neighbourhood of saddle points for example, trajectories approach the fixed point most rapidly along the stable manifold, and depart most rapidly along the unstable manifold. Tangent vectors along the stable manifold are contracting (negative Lyapunov exponents), and tangent vectors along the unstable manifold are expanding (positive Lyapunov exponent). Any region of phase space where these two types of behaviour are in close proximity may exhibit stretching and folding.

Example. Consider a damped, unforced pendulum

$$\frac{\mathrm{d}^2x}{\mathrm{d}t^2} + a\frac{\mathrm{d}x}{\mathrm{d}t} + \sin x = 0.$$

After an analysis as described in Chap. 2 one arrives at the phase plane shown in Fig. 4.5. Stable and unstable manifolds of saddle fixed point $x = \pm\Pi$, $\mathrm{d}x/\mathrm{d}t = 0$ are the trajectories that approach and depart most quickly from unstable fixed points.

Now consider the slightly driven pendulum

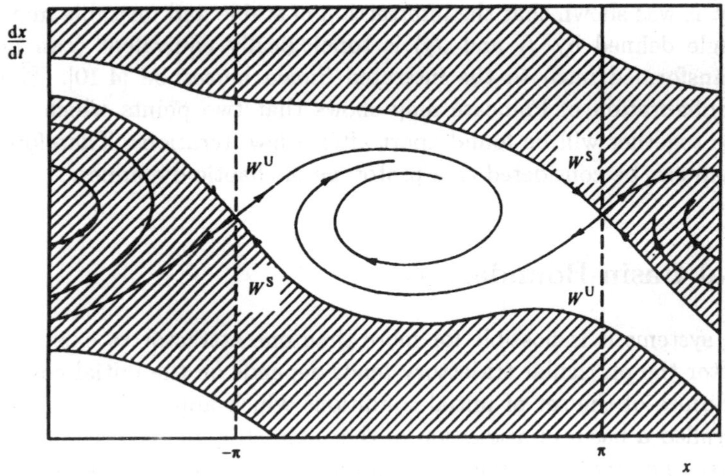

Fig. 4.5. Phase plane of the damped, unforced pendulum

$$\frac{d^2x}{dt^2} + a\frac{dx}{dt} + \sin x = f\cos t,$$

where $f << 1$. In this case the plot of Fig. 4.5 can be considered as a Poincaré map of the three-dimensional phase space, except that the lines should be regarded as a sequence of dots corresponding to successive passages of the trajectories through the cross-section. When forcing is increased, the stable and unstable manifolds can cross first at the point J_1. We can observe that each crossing is mapped into another one closer to the saddle point, leading to an infinite number of intersections $J_2, J_3, \ldots$ (Fig. 4.6). A small rectangular section of the Poincaré map near J_1 suffers stretching and folding much like that of the horseshoe map, due to the strong bending of the manifolds near the saddle point. □

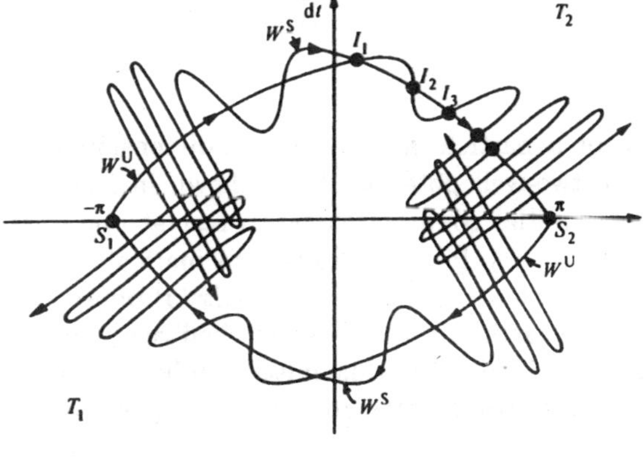

Fig. 4.6. Intersection of stable W^s and unstable W^u manifolds on the Poincaré map of the forced pendulum

Generally, it was shown that the intersection of stable and unstable manifold (rectangle defined by J_1 and J_2) is topologically equivalent (can be smoothly transformed into) to the interated Smale horseshoe [4.10]. This horseshoe configuration of Poincaré map shows that two points which are initially close together will be found apart after a few iterations. Therefore, horseshoe map can be considered as a prototype of chaotic dynamics.

4.5 Fractal Basin Boundaries

In nonlinear systems it is possible for more than one attractor to exist. To which attractor the particular trajectory tends depends on the initial conditions. The closure of the set of initial conditions which approaches a given attractor is called a basin of attraction of that attractor. In the case of two or more co-existing attractors, the transition from one basin of attraction to another is called a basin boundary (see Fig. 4.7).

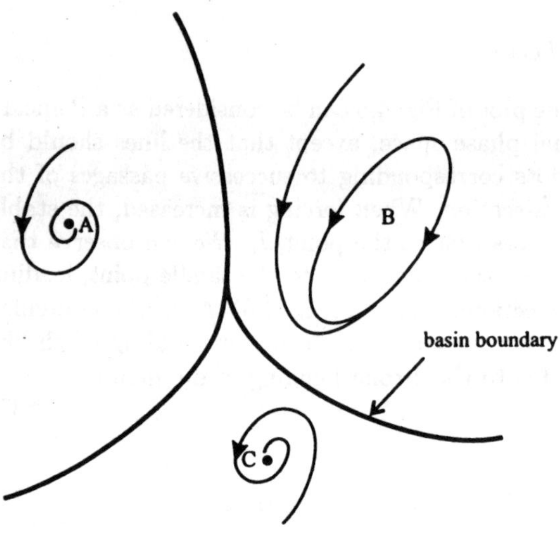

Fig. 4.7. Sketch of basin boundaries

Example. Consider the case of a particle moving in one-dimension under the action of friction and the two-well potential $V(x)$ shown in Fig. 4.8a. Trajectories from almost every initial condition come to rest at one of the stable fixed points $x = x_0$ or $x = -x_0$. In Fig. 4.8b we schematically shown the basins of attraction for these two attractors in the $x - dx/dt$ phase plane of the system. Initial conditions starting in the cross-hatched region are attracted to the attractor at $x = x_0$, $dx/dt = 0$, while initial conditions starting in the unmarked region are attracted to the attractor at $x = -x_0$, $dx/dt = 0$.

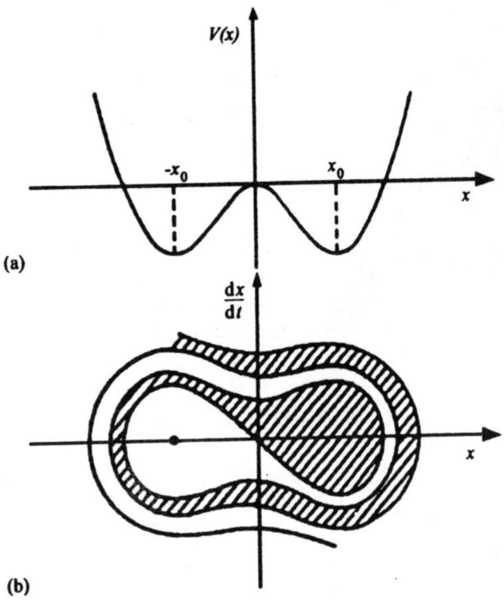

Fig. 4.8. Behaviour of a particle under the action of friction in a two-well potential. (a) Potential $V(x)$ for a point particle moving in one dimension. (b) The basins of attraction for the attractors $x = x_0$ (crosshatched) and at $x = -x_0$ (not crosshatched)

Basins of attraction are separated by a simple curve (the basin boundary). This curve goes through the unstable fixed point $x = dx/dt = 0$. Initial conditions on the basin boundary generate a trajectory that eventually approaches the unstable fixed point, i.e., the basin boundary is the stable manifold of an unstable fixed point. □

Example. Consider the resonance curve of Duffing's equation

$$\frac{d^2y}{dt^2} + 0.175\frac{dy}{dt} + y - 0.01y^3 = \sin\omega t \qquad (4.1)$$

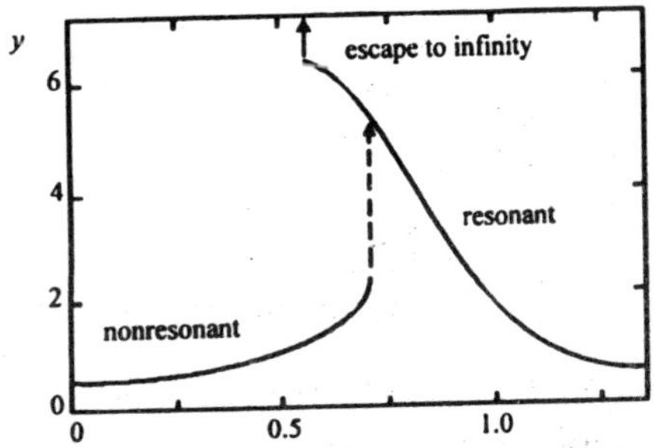

Fig. 4.9. Resonance curve of oscillator (4.1)

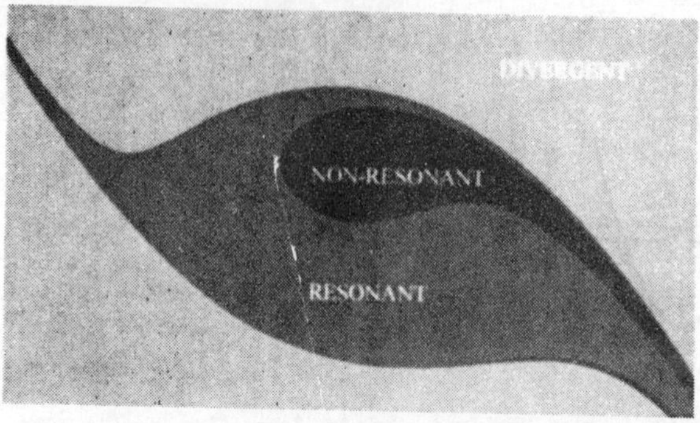

Fig. 4.10. Basin boundaries of oscillator (4.1) for $\omega = 0.825$

shown in Fig. 4.9. Three types of oscillations are possible for this system:

(a) oscillations with amplitude equal to that of the upper branch (resonant),
(b) oscillations with amplitude equal to that of the lower branch (non-resonant), and
(c) oscillations with amplitude diverging to infinity (divergent).

The domains of attraction for these attractors are shown in Fig. 4.10. □

In these examples as well as in the sketch of Fig. 4.7 the basin boundaries are smooth, continuous lines. This implies that when the initial conditions are away from the boundary, small uncertainties in them will not affect the response. However, it has been shown that in many nonlinear systems, this boundary is not smooth and has a fractal structure, so it is called a fractal basin boundary. Of course in this case any small uncertainties in initial conditions may lead to uncertainties in the outcome of the system.

Example. Figure 4.11 shows the basin structure of the forced damped pendulum

$$\frac{\mathrm{d}^2x}{\mathrm{d}t^2} + 0.1\frac{\mathrm{d}x}{\mathrm{d}t} + \sin x = 2\cos t.$$

In this case there are two periodic attractors that have the same period as the forcing. The orbit for one of these attractors has an average clockwise motion (negative average value of $\mathrm{d}x/\mathrm{d}t$), while the orbit for the other attractor has an average counterclockwise motion.

In Fig. 4.11 the black region represents initial values of x and $\mathrm{d}x/\mathrm{d}t$ that tend to the attractor whose orbit has average counterclockwise motion, while the white region represents initial values that generate clockwise motion. We can see that there is a small scaled structure on which the black and white regions appear to be finely interwoven, i.e., the basin boundary shows a fractal nature [4.4]. □

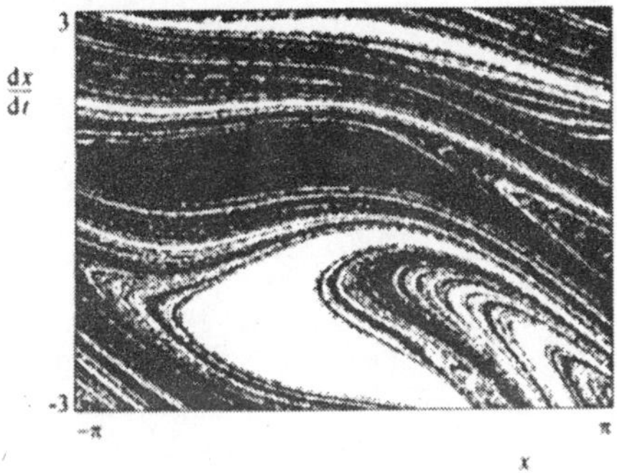

Fig. 4.11. Fractal basin boundary of pendulum

The existence of fractal basin boundaries shows that even a system with periodic response can show a sensitive dependence on the initial conditions in the sense that trajectories of nearby initial conditions converge to different periodic attractors.

Problems

1. What is the capacity dimension of the Cantor set obtained by removing the middle interval of length
   (a) one half
   (b) one quarter
   (instead one third as in Fig. 4.1) of the intervals on the previous stage of construction.
2. Estimate the capacity dimension of the set
$$S = [1, 1/2, 1/3, 1/4, ...]$$
3. Show that the Henon map
$$x_{n+1} = 1 - ax_n^2 + y_n$$
$$y_{n+1} = bx_n$$
   can produce horseshoes.
4. Estimate the basin boundaries of co-existing attractors in
   (a) the damped pendulum
$$\frac{d^2x}{dt^2} + a\frac{dx}{dt} + \sin x = 0,$$
   (b) the Helmholtz oscillator
$$\frac{d^2x}{dt^2} + \frac{dx}{dt} - x - x^2 = 0.$$

5. Routes to Chaos

In the previous chapters we have introduced the methods for describing chaotic behaviour. Here we will observe how the behaviour of our systems changes during the transition from periodic to chaotic states. The mechanism of the transition to chaos is of fundamental importance for understanding the phenomenon of chaotic behaviour. There are three main routes to chaos which can be observed in nonlinear oscillators.

5.1 Period-Doubling

We have already discussed the period-doubling route to chaos for the logistic equation. Here we consider driven anharmonic oscillators. Let us observe the oscillations of Duffing's oscillator

$$\frac{\mathrm{d}^2x}{\mathrm{d}t^2}+\frac{\mathrm{d}x}{\mathrm{d}t}-10x+100x^3=F\cos(3.5t)\ , \tag{5.1}$$

where F is the amplitude of the excitation force and $\Omega = 3.5$ its frequency. By changing a control parameter F it is possible to obtain the phase portraits and Poincaré maps presented in Fig. 5.1a–d.

We can observe that increasing one of the parameters first we observe one point in Poincaré map (Fig. 5.1a), next two points (Fig. 5.1b), four points (Fig. 5.1c) and finally an infinite number of points (Fig. 5.1d). As has been described in Chap. 2, one point in Poincaré maps of nonautonomous systems like (5.1) means that the period of beam oscillations is equal to a period of the excitation force $T = 2\pi/\Omega$, while two points indicate that the period of oscillations is twice as long. The transition to oscillations with twice the period is called a period-doubling bifurcation. If oscillations with double period are stable we call this type of bifurcation supercritical and when they are unstable the bifurcation is subcritical.

The route to chaos via period-doubling bifurcations has got some very interesting universal properties. In Chap. 3 have already discussed this route for the logistic equation. More examples can be found in Sects. 5.4, 5.5, 6.1 and 6.4.

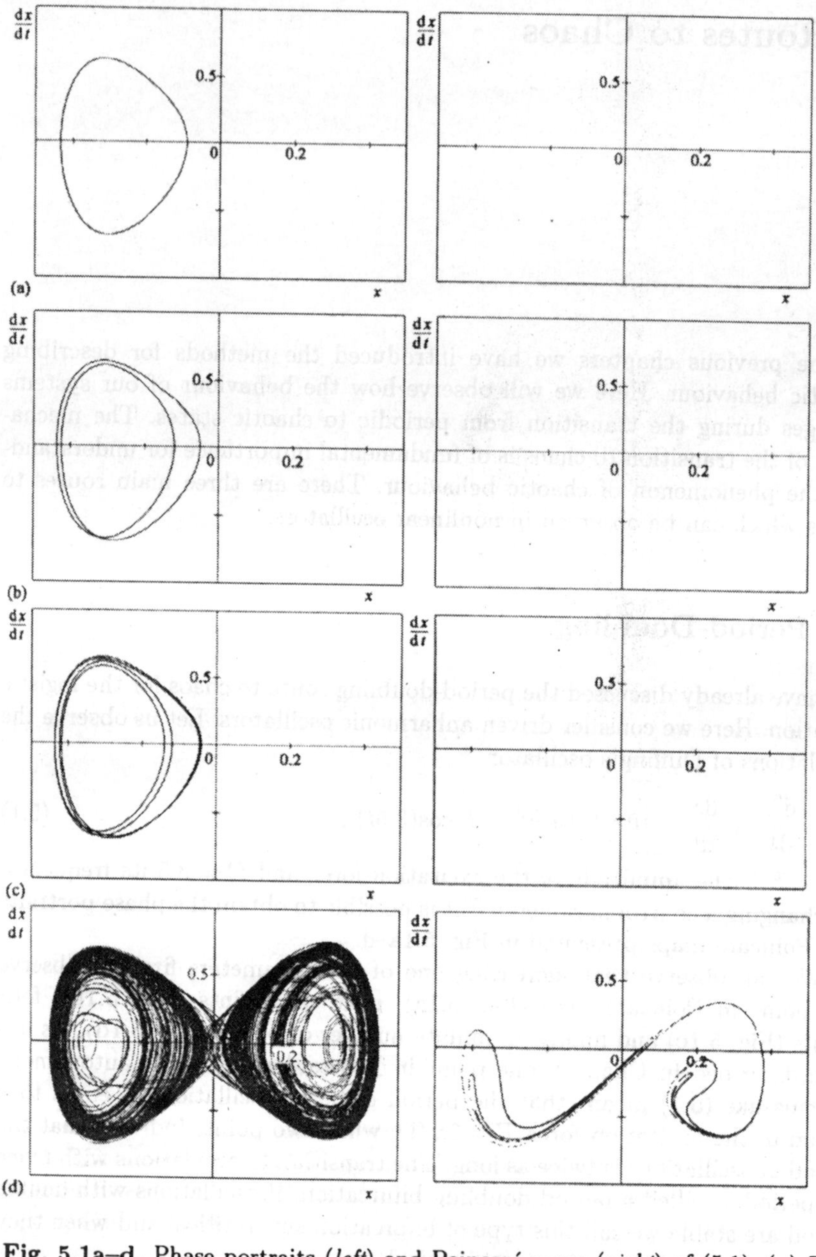

Fig. 5.1a–d. Phase portraits (*left*) and Poincaré maps (*right*) of (5.1). **(a)** T periodic solution, $F = 0.70$, **(b)** $2T$ periodic solution, $F = 0.80$. **(c)** $4T$ periodic solution, $F = 0.82$. **(d)** chaotic solution, $F = 0.85$

5.2 Quasiperiodic Route

The second route to chaos is connected with Hopf bifurcation, which generates a limit cycle starting from a fixed point. As an example of Hopf bifurcation consider the following equation:

$$\frac{\mathrm{d}z}{\mathrm{d}t} = (\sigma + \mathrm{i}\omega)z - g|z|^2 z, \tag{5.2}$$

where z is complex and

$$z(t) = x(t) + \mathrm{i}y(t)$$

with x and y real. Furthermore ω is real and positive and g is complex. The bifurcation takes place when the control parameter σ goes through 0 from negative values. Let $g = g' + \mathrm{i}g''$ and insert

$$z = |z|\exp(\mathrm{i}\phi)$$

in (5.2). This gives

$$\exp(\mathrm{i}\phi)\left\{\frac{\mathrm{d}|z|}{\mathrm{d}t} + \mathrm{i}|z|\frac{\mathrm{d}\phi}{\mathrm{d}t}\right\} = \left\{(\sigma + \mathrm{i}\omega)|z| - (g' + ig'')|z|^3\right\}.$$

The time evolution of the modulus $|z|$ and phase ϕ are therefore given by

$$\frac{\mathrm{d}|z|}{\mathrm{d}t} = (\sigma - g'|z|^2)|z|, \tag{5.3}$$

$$\frac{\mathrm{d}\phi}{\mathrm{d}t} = \omega - g''|z|^2 \tag{5.4}$$

except at the fixed point $|z| = 0$. One can see from (5.3) that the modulus is not coupled with the phase so that its evolution can be solved independently. Fixed points for (5.3) are given by

$$|z|_\mathrm{f} = 0 \quad \text{and} \quad |z|_\mathrm{f} = \sqrt{\frac{\sigma}{g'}}. \tag{5.5}$$

The Hopf bifurcation is called supercritical when $g' > 0$ since a nontrivial solution exists for $\alpha > 0$, i.e. above the linear threshold. When $g' < 0$, the nontrivial solution exists for $\sigma < 0$, i.e. below the threshold and the bifurcation is called ***subcritical.*** The stability of the bifurcation solutions is analysed by inserting $|z| = |z|_\mathrm{f} + |z|'|$ in (5.3) with $|z|_\mathrm{f}$ given by (5.5), which yields

$$\frac{\mathrm{d}|z|'}{\mathrm{d}t} = \sigma(|z|_\mathrm{f} + |z|') - g'(|z|_\mathrm{f} + |z|')^3.$$

It follows that

$$\frac{\mathrm{d}|z|}{\mathrm{d}t} = \sigma|z|_\mathrm{f} - g'|z|_\mathrm{f}^3 + (\sigma - 3g'|z|_\mathrm{f}^2)|z|' + \ldots$$

Terms independent of $|z|'$ cancel exactly by virtue of the fixed point condition, which can also be used to simplify the expression of the coefficient of $|z|'$, so that we simply get

$$\frac{\mathrm{d}|z|'}{\mathrm{d}t} = -2\sigma|z|'.$$

Therefore, when the bifurcation is supercritical, fixed points exist for $\sigma > 0$ and perturbations are damped as $|z|' \exp(-2\sigma)$ and $-2\sigma < 0$. In the opposite case (subcritical bifurcation) the nontrivial fixed points are unstable. The trivial fixed point $|z|_f = 0$ is stable for $\sigma < 0$ and unstable for $\sigma > 0$. If $|z|$ is known, the time evolution of the phase ϕ can be determined. Let us consider simply the asymptotic regime where $|z|$ has reached the fixed point value. We have

$$\frac{\mathrm{d}\phi}{\mathrm{d}t} = \omega - \frac{g''\sigma}{g'},$$

which can be solved as

$$\phi(t) = \phi_0 + (\omega - \frac{g''\sigma}{g'})t. \tag{5.6}$$

Equation (5.6) shows that as the result of supercritical Hopf bifurcation we observe the birth of the limit cycle i.e. a periodic motion with a frequency slightly corrected from the nonlinear contribution. The birth of the limit cycle is illustrated in Fig. 5.2 and the bifurcation diagram of (5.6) is given in Fig. 5.3.

In many systems undergoing Hopf bifurcation, after further increase of the control parameter it is possible to find the secondary Hopf bifurcation. After this bifurcation the response of the system is quasiperiodic with two independent frequencies. The Poincaré map defined by the frequency of the

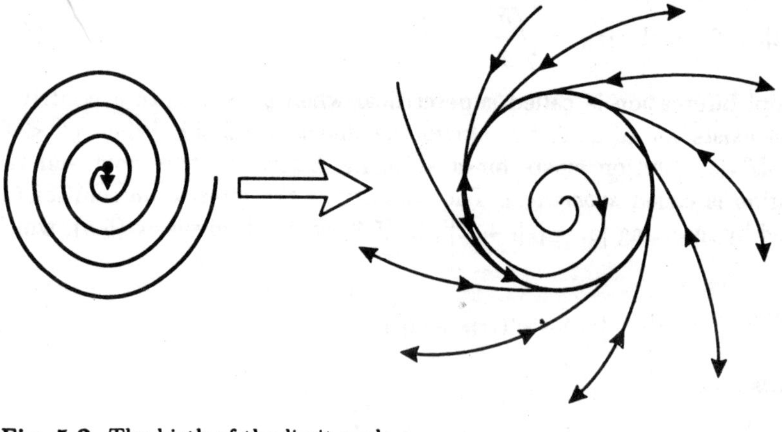

Fig. 5.2. The birth of the limit cycle

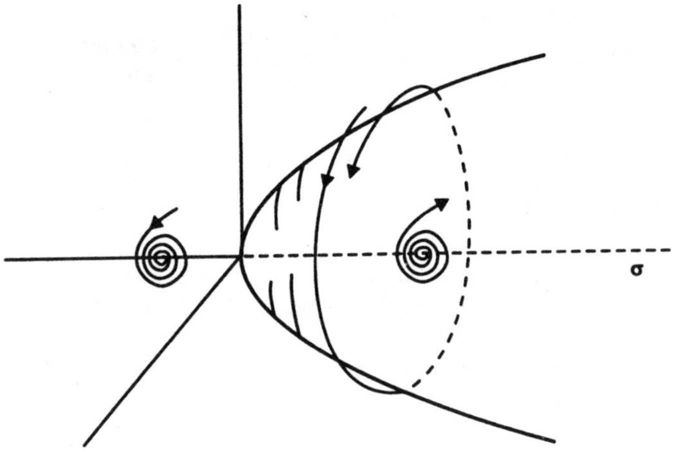

Fig. 5.3. Bifurcation diagram of (5.6)

limit cycle obtained by the first Hopf bifurcation before secondary Hopf bifurcation has a fixed point, which after bifurcation evolves to a limit cycle with another frequency, so that trajectories are drawn on a two-dimensional torus T^2.

In 1944 Landau [5.2] first presented the following route to turbulence in time. In this hypothesis the chaotic state is approached by an infinite sequence of Hopf bifurcations as shown in Fig. 5.4.

The main disadvantage of this model is that the power spectrum always remains discrete, whereas for chaotic behaviour we have a continuous spectrum.

A modification of this model, which better describes the route observed in many numerical and experimental researches, was proposed by Newhouse, Ruelle, and Takens [5.1]. They suggested a much shorter route than that proposed by Landau [5.2]. They showed that after three Hopf bifurcations regular motion becomes highly unstable in favour of motion on a strange attractor as presented in Fig. 5.5.

We finish this section with the Hopf theorem.

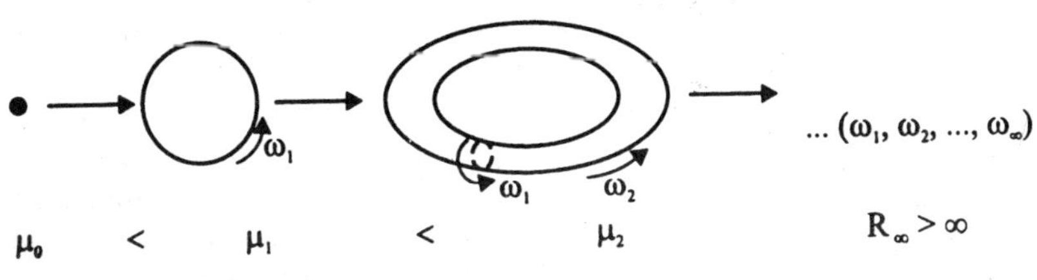

Fig. 5.4. Landau's route to chaos

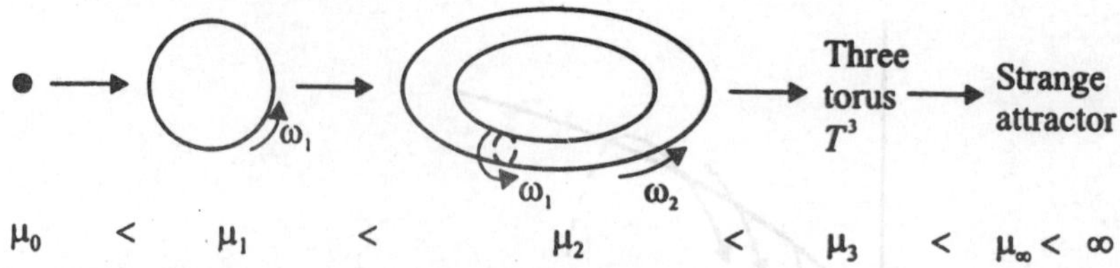

Fig. 5.5. Newhouse–Ruelle–Takens route to chaos

Theorem. Let G be an open connected domain in $\mathcal{R}^n$, $c > 0$, and let F be a real analytic function defined on $G \times [-c, c]$. Consider the differential system:

$$\frac{\mathrm{d}x}{\mathrm{d}t} = F(x, \mu), \quad \text{where} \quad x \in G, \ |\mu| < c. \tag{5.7}$$

Suppose there is an analytic, real, vector function g defined on $[-c, c]$ such that $F(g(\mu), \mu) = 0$. Thus one can expand $F(x, \mu)$ about $g(\mu)$ in the form

$$F(x, \mu) = L_\mu \bar{x} + F^*(\bar{x}, \mu), \quad \bar{x} = x - g(\mu), \tag{5.8}$$

where L_μ is an $n \times n$ real matrix which depends only on μ, and $F^*(\bar{x}, \mu)$ is the nonlinear part of F. Suppose there exist exactly two complex conjugate eigenvalues $\alpha(\mu)$, $\bar{\alpha}(\mu)$ of L_μ with the properties

$$\mathrm{Re}(\alpha(0)) = 0 \quad \text{and} \quad \mathrm{Re}(\alpha'(0)) \neq 0 \quad ('= \mathrm{d}/\mathrm{d}\mu).$$

Then there exists a periodic solution $P(t, \epsilon)$ with period $T(\epsilon)$ of (5.7) with $\mu = \mu(\epsilon)$, such that $\mu(0) = 0$, $P(t, 0) = g(0)$ and $P(t, \epsilon) \neq g(\mu(\epsilon))$ for all sufficiently small $\epsilon \neq 0$. Moreover $\mu(\epsilon)$, $P(t, \epsilon)$, and $T(\epsilon)$ are analytic at $\epsilon = 0$, and $T(0) = 2\pi/|\mathrm{Im}\ \alpha(0)|$. These 'small' periodic solutions exist for exactly one of three cases: either only for $\mu > 0$, or only for $\mu < 0$, or only for $\mu = 0$.

5.3 Intermittency

By intermittency we mean the occurrence of fluctuations that alternate 'randomly' between long periods of regular behaviour and relatively short irregular bursts, i.e. the motion is nearly periodic with occasional irregular bursts. Typical time histories of the intermittent transitions are shown in Fig. 5.6.

It has been found that the density of chaotic bursts increases with an external control parameter, which shows that intermittency presents a continuous route from regular to chaotic behaviour. The theory of intermittency has been established in a pioneering study of Pomeau and Manneville [5.3]. To understand the mechanism of intermittency let us go back to the logistic map (3.6). As has been pointed out earlier the regions of chaotic behaviour are interrupted by intervals of periodic behaviour (the so-called windows). One of the largest of these windows occurs near $a = 3.83$, where a periodic orbit (a 3-cycle) occurs. Let us consider the third return map:

$$x_{n+3} = f\{f[f(a, x_n)]\}. \tag{5.9}$$

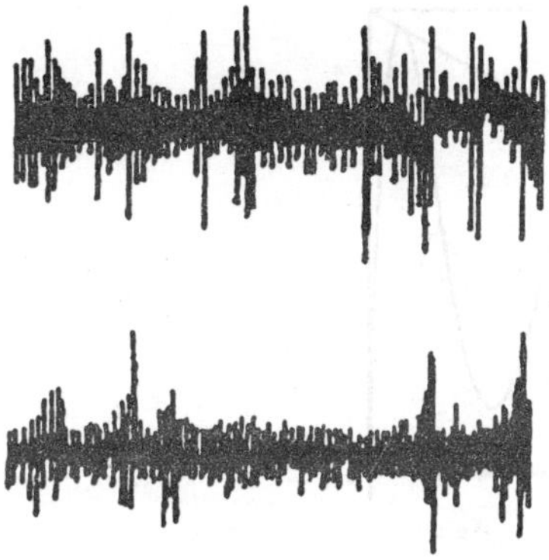

Fig. 5.6. Typical time series of intermittent behaviour

In Fig. 5.7 three such maps are shown for three values of a: (a) shortly before the window, (b) at the start of the window and (c) inside the window. At the beginning of the window the third-order return map shows three values of x where the curve is tangent to the diagonal line $x_{n+3} = x_n$. These points are the cyclic steady state values of x, i.e. other initial values of x will be drawn to these fixed points since the shallow slopes of the curve near the fixed points indicate stability. This particular event which occurs for $a = a_{\mathrm{T}}$ is called a tangent bifurcation. For a slightly larger than a_{T} the curve of the corresponding third-order return map crosses the diagonal at three pairs of values of x (the other crossing is unstable). It can be shown that the magnitudes of the slopes for three of these points (one in each pair) are larger than 1, while for the other three they are less than 1. The first ones are repellers while the second ones are attractors and therefore the cyclic behaviour initiated by the tangent bifurcation continues to be stable.

An interesting situation occurs for a just below the onset of the period 3 window, the third order return map is not quite tangent to the diagonal line. Therefore x_n can pass through the resulting narrow gaps and then go freely around the plane until it again becomes temporarily trapped in a narrow gap as shown in Fig. 5.7d, while in the outside gap we have an irregular burst characteristic for intermittency.

In this type of intermittency one can easily check that the map has got real eigenvalues; fixed points are stable when eigenvalues are less than 1 and unstable if they are larger than 1. This type of intermittency is called the first type. It is characterised by the loss of stability when the real eigenvalue crosses the unit circle at + 1. The second type is characterised by the simultaneous crossing of the two complex eigenvalues of the unit circle, i.e. the system undergoes subcritical Hopf bifurcation. When the real eigenvalue crosses the unit circle at -1, i.e. the system undergoes a subcritical period-doubling bifurcation, we have the third type of intermittency which is described in Problem 4.

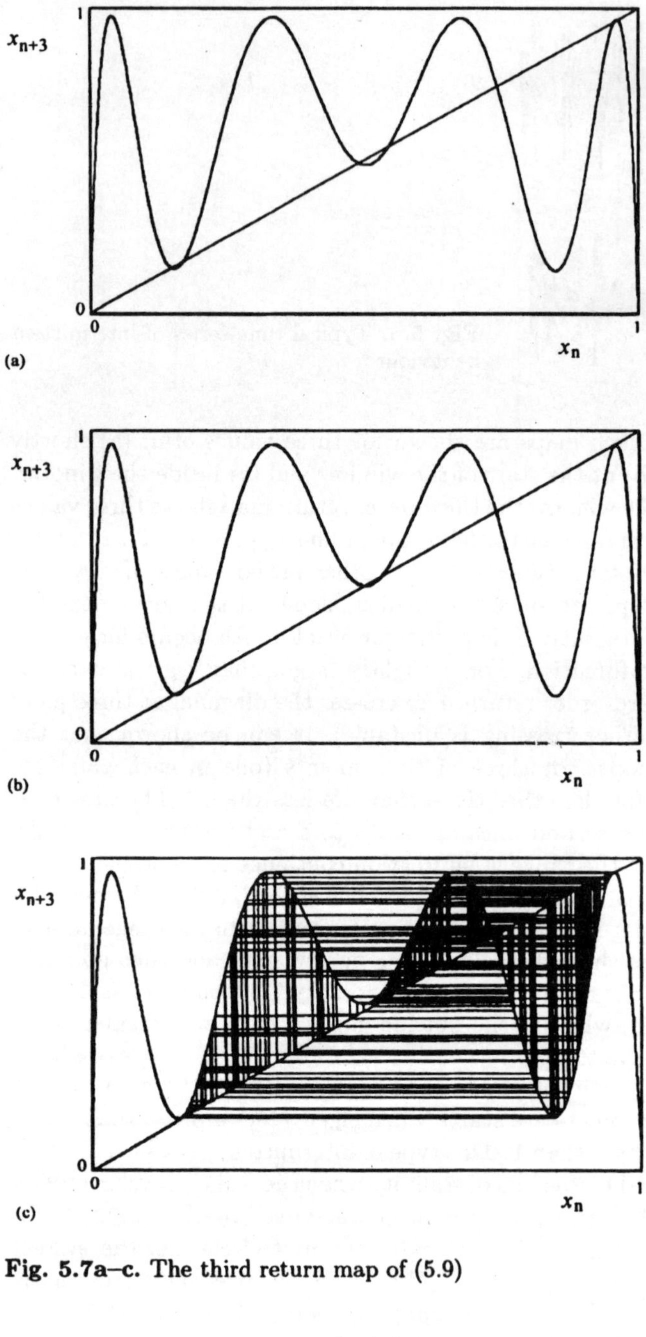

Fig. 5.7a–c. The third return map of (5.9)

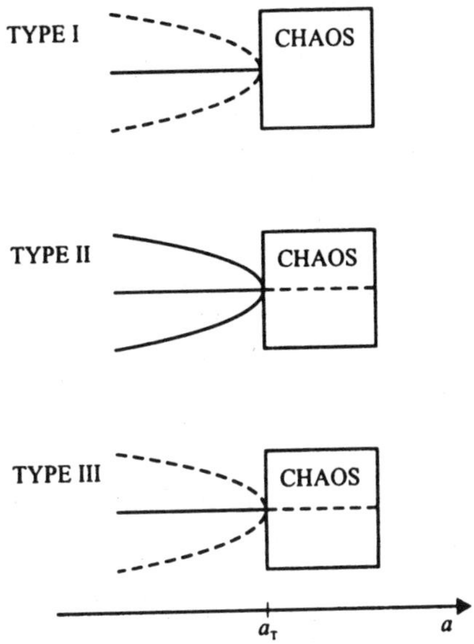

Fig. 5.8. Three types of intermittency

Each of the three types of intermittency transitions corresponds to one of the three types of generic bifurcations described in Chap. 3,

Type I: tangent (or saddle-node),

Type II: Hopf,

Type III: subcritical period-doubling.

These bifurcations are shown in Fig. 5.8 where solid lines indicate stable, and dashed lines unstable solutions.

5.4 Duffing's Oscillator: Discrete Dynamics Approach

As was seen in Chap. 3, the maps of discrete-time dynamical systems maps are easier to investigate than typical continuous-time systems. The property has been used as a main idea in the introduction of Poincaré maps. Although in the general case it is rather difficult to give analytical description of Poincaré maps in this section, we consider a map on the interval which has some properties of Duffing's system:

$$\frac{dx^2}{dt^2} + k\frac{dx}{dt} + x^{2v+1} = b\cos t, \tag{5.10}$$

where $v = 2, 3, 4, \ldots$. If k is small and B is large, it is possible to obtain a one-dimensional approximation of the Poincaré map from the maximum absolute point of x [5.4].

Let $p = (t, x, \dot{x})$ be a solution of (5.10). The maximum absolute point is $p = (t, x, \dot{x})$ in $x > 0$ for $\mathrm{d}^2x/\mathrm{d}t^2 < 0$ and in $x < 0$ for $\mathrm{d}^2x/\mathrm{d}t^2 > 0$. If

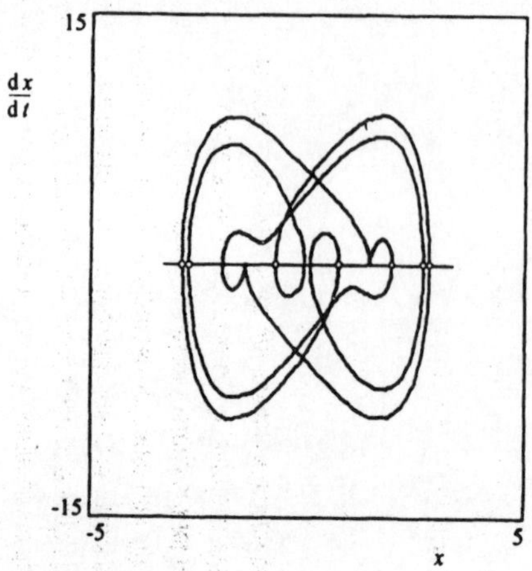

Fig. 5.9. Maximum absolute points

we let $p_n(t_n, x_n, 0)$, $p_{n+1}(t_{n+1}, x_{n+1}, 0)$ be two successive maximum absolute points, then one obtains a discrete dynamical system:

$$T : p_n \to p_{n+1}. \tag{5.11}$$

Examples of maximum absolute points are shown in Fig. 5.9. When the interval $t_n \le t \le t_{n+1}$, where $p_n = (t_n, x_n, 0)$ is a maximum absolute point, is small, one can introduce the piecewise linearised version (variational) of (5.10) in the form

$$\frac{\mathrm{d}^2 x}{\mathrm{d}t^2} + k\frac{\mathrm{d}x}{\mathrm{d}t} + a(x_n)x = B\cos t, \tag{5.12}$$

where $a(x_n) = (2v+1)x^{2v}$. Equation (5.12) is valid when the interval (t_n, t_{n+1}) is sufficiently small, i.e. B is sufficiently large and x_n is large. If $a(x_n) > k^2/2$ which also holds in Ueda's system, the solution of the piecewise linear equation is

$$x(t) = A\mathrm{e}^{-\frac{k}{2}t}\cos(\omega_n t + \delta_1) + \frac{B}{[(a-1)^2 + k^2]^{\frac{1}{2}}}\cos(t + \delta_2), \tag{5.13}$$

where A and δ_1 are determined by the initial conditions and

$$\delta_2 = \tan^{-1}\frac{k}{a-1}, \quad \omega_n = \left[a(x_n) - k^2/4\right]^{\frac{1}{2}}. \tag{5.14}$$

From the definition of map t, the initial conditions are $x(t_n) = x_n$, $\mathrm{d}x(t_n)/\mathrm{d}t = 0$ and one obtains the map T in the form:

$$\begin{aligned} \theta_{n+1} &\approx \theta_n + 1/\omega_n, \\ x_{n+1} &\approx x(2\pi\theta_n + 2\pi/\omega_n), \end{aligned} \tag{5.15}$$

where $\theta_n = t_n/2\pi \pmod 1$, and can be identified as the phase of the periodic perturbation. When the damping is small $k \ll 1$, and $B \gg 1, x_n \gg 1$, then we have

$$\theta_{n+1} \approx \theta_n + \frac{1}{|x_n|^v (2v+1)^{\frac{1}{2}}},$$

$$x_{n+1} \approx x_n - \frac{2\pi B}{(2v+1)^{\frac{3}{2}} |x_n|^{3v}} \sin 2\pi\theta_n. \tag{5.16}$$

We can approximate the recursion relations (5.16) by differential equations in which the step space is $d\tau$,

$$\frac{d\theta}{d\tau} \approx \frac{1}{(2v+1)^{\frac{1}{2}} |x|^v},$$

$$\frac{dx}{d\tau} \approx \frac{2\pi B}{(2v+1)^{\frac{1}{2}} |x|^{3v}} \sin 2\pi\theta. \tag{5.17}$$

These equations are readily integrated,

$$x \approx (B \cos 2\pi\theta)^{\frac{1}{2v+1}}, \tag{5.18}$$

which was found to be in good agreement with computer simulations near the bifurcation point.

Equation (5.18) allows one to obtain the one-dimensional Poincaré map in the following form:

$$\theta_{n+1} = f(\theta_n) \approx \theta_n + [1/(2v+1)^{\frac{1}{2}}](|B \cos 2\pi\theta_n|)^{-v/(2v+1)}. \tag{5.19}$$

The consideration of the dynamics of Poincaré map (5.15) is limited to the interval $[0,1]$ so the map (5.19) has to be understood as

$$\theta_{n+1} = f(\theta_n) - [f(\theta_n)], \tag{5.20}$$

where $[f(\theta_n)]$ indicates the rational part of $f(\theta_n)$. This map acts on the interval $[0,1] - 0.25, 0.75$. The bifurcation diagram of the above map for $v = 1$ is shown in Fig. 5.10, where the bifurcation parameter

$$A = \frac{1}{\sqrt{3}|B|^{\frac{1}{3}}}$$

has been taken. This diagram shows the very complicated behaviour of the map, showing also that it is worth considering its behaviour not only in connection with Duffing's oscillator, but also to concentrate, as here, on the behaviour of the map for parameters close to the transition to chaos in Ueda's system. The enlarged part of this diagram for $B \in [8.77, 18.7]$ is shown in Fig. 5.11a–c.

Now let us consider a scaling law for our map. If there is an m-periodic orbit of f then the dependence of m on B is obtained by applying a theory of intermittency. Approximating the discrete dynamical equation (5.19) again by a differential equation

$$d\theta/d\tau \approx (2v+1)^{-\frac{1}{2}} (|B \cos 2\pi\theta|)^{-v/(2v+1)}$$

and integrating it we obtain as a period

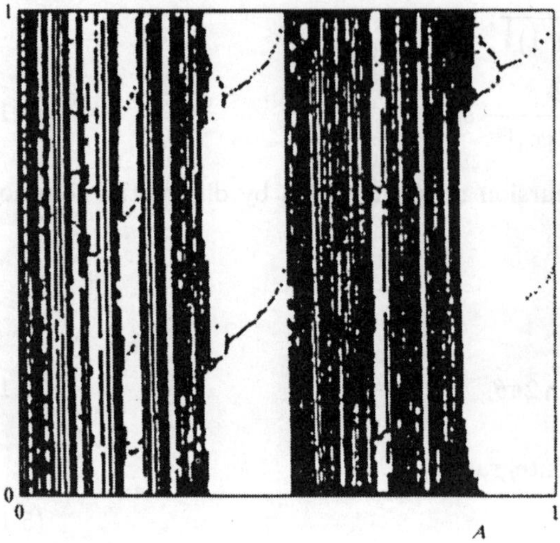

Fig. 5.10. The bifurcation diagram of the map (5.20)

$$T \equiv \int_0^1 \mathrm{d}\tau = B^{v/(2v+1)}(2v+1)^{\frac{1}{2}} \int_0^1 |\cos 2\pi\theta|^{v/(2v+1)} \mathrm{d}\theta = C(v) B^{v/(2v+1)},$$

where $C(v)$ represents a B-independent constant. Since $T \propto m$ holds roughly in the m-periodic state the scaling law is

$$B_m \propto m^{(2v+1)/v}.$$

The comparison of the theoretical $\alpha_t \equiv (2v+1)v$ and experimental results of Duffing's oscillator is shown in Table 5.1.

Table 5.1. Comparison of the theoretical and experimental scaling laws

v	α_{exp}	α_t
1	3.18	3.00
2	2.57	2.50
3	2.42	2.33

5.5 Condition for Chaos by Period Doubling Route

In Chap. 1 we introduced the harmonic balance method and its application to investigate the response of a Duffing oscillator. Now we show how this method can be used to estimate the region of chaotic behaviour when this state is obtained by a period-doubling bifurcation. The accumulation points

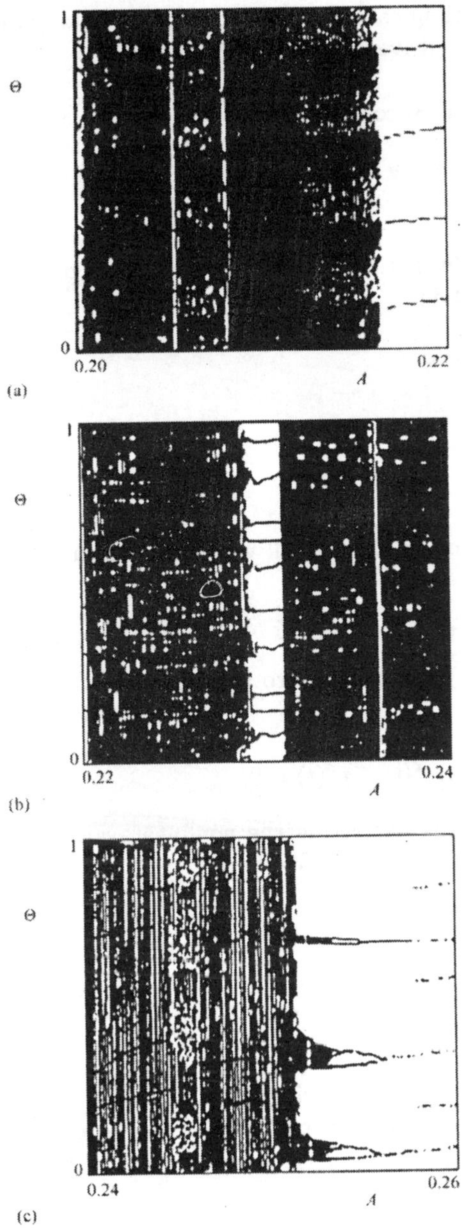

Fig. 5.11a–c. Details of the bifurcation diagram of Fig. 5.10

of stable and unstable period-doubling cascades are proposed as boundaries of the chaotic domain. Consider the driven Duffing's equation

$$\frac{d^2x}{dt^2} + a\frac{dx}{dt} + bx + cx^3 = B_0 + B_1 \cos(\Omega t) \tag{5.21}$$

and the first approximate solution in the form

$$x(t) = C_0 + C_1 \cos(\Omega t + \zeta), \tag{5.22}$$

where C_0, C_1 and ζ are constants. Substituting (5.23) into (5.22) it is possible to determine these constants by solving

$$-C_1\Omega^2 + \frac{3}{4}C_1^3 + 3C_0^2C_1 = B_1 \cos\zeta,$$

$$-aC_1\Omega = B \sin\zeta,$$

$$C_0^2 + \frac{3}{2}C_0C_1^2 = B_0. \tag{5.23}$$

To study the stability of the solution (5.23) a small variational term $y(t)$ is added to it

$$x(t) = C_0 + C_1 \cos(\Omega t + \zeta) + y(t). \tag{5.24}$$

After inserting (5.24) into (5.22) and taking into account (5.31) one obtains the following variational term

$$\frac{d^2y}{dt^2} + a\frac{dy}{dt} + 3x^3(t)y(t) + 3x(t)y^2(t) + y^3(t) = 0. \tag{5.25}$$

The local stability is examined by neglecting the nonlinear variational terms in (5.25) then inserting (5.23) into (5.25). We have

$$\frac{d^2y}{dt^2} + a\frac{dy}{dt} + y[\lambda_0 + \lambda_1 \cos\theta + \lambda_2 \cos 2\theta] = 0, \tag{5.26}$$

where

$$\lambda_0 = 3C_0^2 + (3/2)C_1^2, \quad \lambda_1 = 6C_0C_1, \quad \lambda_2 = \frac{3}{2}C_1^2, \quad \theta - \Omega t + \zeta. \tag{5.27}$$

In the derivation of (5.26), it was assumed for simplicity but without loss of generality that $B_0 = 0$. As we have a parametric term of frequency $\Omega - \lambda_1 \cos\theta$, the lowest order unstable region is that which occurs close to $\Omega/2 \approx \sqrt{\lambda_0}$ and at its boundary we have the solution

$$y(t) = b_{1/2} \cos[(\Omega/2)t + \rho]. \tag{5.28}$$

To determine boundaries of the unstable region we insert (5.28) into (5.26), and the conditions of nonzero solution for $b_{1/2}$ lead us to the following criterion to be satisfied at the boundary

$$(\lambda_0 - \Omega^2/4)^2 + a^2\Omega^2/4 - \lambda_1^2/4 = 0. \tag{5.29}$$

From (5.29) one obtains the interval $(\Omega_1^{(2)}, \Omega_2^{(2)})$ within which period 2 solutions exist. Further analysis (for more details see [5.5]) show that at Ω_2 we have a stable period-doubling bifurcation for decreasing Ω, and at Ω_1 an unstable period-doubling bifurcation for increasing Ω.

In this interval we can consider the period 2 solution of the form

$$x(t) = A_0 + A_{1/2}\cos((\Omega/2)t + \rho) + A_1 \cos \Omega t, \tag{5.30}$$

where A_0, $A_{1/2}$, A_1 and ρ are constants to be determined. Again, to study the stability of the period 2 solution, we have to consider a small variational term $y(t)$ added to (5.30). After linearisation one obtains

$$\frac{d^2y}{dt^2} + a\frac{dy}{dt} + y\Bigg[\lambda_1^{(2)} + \lambda_{1/2c}\cos\frac{\Omega}{2}t + \lambda_{1/2s}\sin\frac{\Omega}{2}t$$
$$+\lambda_{3/2}\cos\left(\frac{3\Omega}{2}t+\rho\right) + \lambda_{1c}^{(2)}\cos\Omega t + \lambda_{1s}^{(2)}\sin\Omega t$$
$$+\lambda_2^{(2)}\cos 2\Omega t\Bigg] = 0, \tag{5.31}$$

where

$$\lambda_0^{(2)} = 3(A_0^2 + 0\cdot 5A_{1/2}^2 + 0\cdot 5A_1^2),$$
$$\lambda_{1/2c} = 3A_{1/2}(2A_0 + A_1)\cos\rho,$$
$$\lambda_{1/2s} = 3A_{1/2}(A_1 - 2A_0)\sin\rho,$$
$$\lambda_{3/2} = 3A_1A_{12},$$
$$\lambda_{1c}^{(2)} = 6A_0A_1 + (3/2)A_{1/2c}^2\cos 2\rho,$$
$$\lambda_{1s}^{(2)} = -(3/2)A_{1/2s}^2\sin 2\rho,$$
$$\lambda_2^{(2)} = (3/2)A_1^2.$$

The form of (5.31) enables us to find the range of existence of a period 4 solution represented by

$$y(t) = b_{1/4}\cos\left(\frac{\Omega}{4}t+\rho\right) + b_{3/4}\cos\left(\frac{3\Omega}{4}t+\rho\right). \tag{5.32}$$

After inserting (5.32) into (5.31) the condition of nonzero solution for $b_{1/2}$ and $b_{3/4}$ gives us the following set of nonlinear algebraic for Ω, $\cos\phi$, and $\sin\phi$ to be satisfied for existence:

$$\left(\lambda_{1/2s} + \lambda_{1s}^{(2)}\right) - 0.5\left(\lambda_{1/2c} - \lambda_{1c}^{(2)}\right)\left(-\frac{a\Omega}{2} + \lambda_{1/2s} - \lambda_{3/2}\sin\rho\right) = 0,$$

$$\frac{9}{8}\Omega^2 + 0.5\lambda_0^{(2)} + \lambda_{3/2}\cos\rho - 0.5\left(\lambda_{1/2c} + \lambda_{1c}^{(2)}\right)\left(\lambda_{1s}^{(2)} + \lambda_{1/2c}\right) = 0,$$

$$-\frac{3}{2}a\Omega - \lambda_{3/2}\sin\rho - 0.5\left(\lambda_{1/2c} + \lambda_{1c}^{(2)}\right)\left(\lambda_{1s}^{(2)} + \lambda_{1/2c}\right) = 0. \tag{5.33}$$

Solving (5.33) by a numerical procedure it is possible to obtain $\Omega_1^{(4)}$ and $\Omega_2^{(4)}$, the frequencies of stable and unstable period 4 bifurcations. We assume that Feigenbaum's description of period doubling described in Sect. 3.2 is valid for our system, i.e.

$$\frac{\left(\Omega_{1,2}^{(2^n)} - \Omega_{1,2}^{(2^{n-1})}\right)}{\left(\Omega_{1,2}^{(2^{n+1})} - \Omega_{1,2}^{(2^n)}\right)} \longrightarrow \delta$$

as $n \to \infty$, where $\delta = 4.669$ is the Feigenbaum constant and $n = 1, 2, \ldots$. We note that $\Delta\Omega_{1,2}^{(2^n)}$, $\Delta\Omega_{1,2}^{(2^{n+1})}$ where

$$\Delta\Omega_{1,2}^{(2^n)} := \Omega_{1,2}^{(2^n)} - \Omega_{1,2}^{(2^{n-1})}$$

form an infinite geometrical series with a ratio $1/\delta$. One can estimate limits of both period-doubling cascades as

$$\begin{aligned} \Omega_1^{(\infty)} &= \Omega_1^{(2)} + \Delta\Omega_1/(1 - 1/\delta)\,, \\ \Omega_2^{(\infty)} &= \Omega_2^{(2)} - \Delta\Omega_2/(1 - 1/\delta)\,, \end{aligned} \tag{5.34}$$

where

$$\Delta\Omega_1 = \Omega_1^{(4)} - \Omega_1^{(2)}, \qquad \Delta\Omega_2 = \Omega_2^{(2)} - \Omega_2^{(4)}.$$

The domain where chaotic behaviour can occur is proposed to be between the limits of unstable and stable period-doubling cascades, in the interval $\left(\Omega_1^{(\infty)}, \Omega_2^{(\infty)}\right)$ and of course to expect chaos one must have

$$\Omega_1^{(\infty)} < \Omega_2^{(\infty)}.$$

In Fig. 5.12 we show the comparison of this analytical estimation of the chaotic domain and the actual (numerically found) chaotic domain obtained by Ueda [5.6]. They are in good agreement.

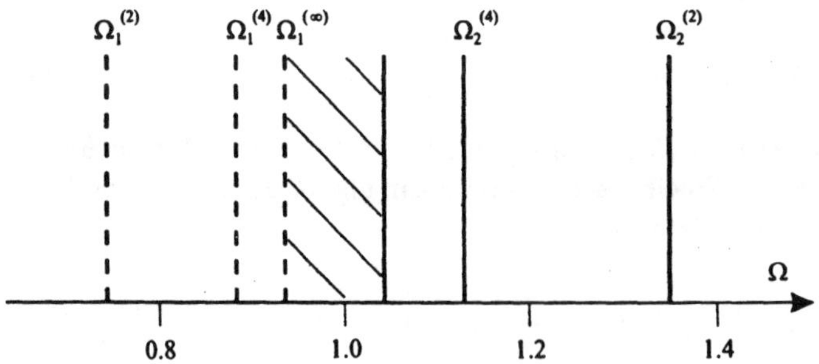

Fig. 5.12. Comparison of analytically and numerically estimated chaotic zones

The analytical technique presented in this section is based on:

(a) approximate period 1, 2 and 4 solutions and their stability limits computed by harmonic balance method,
(b) Feigenbaum's universal constant for the asymptotic ratio of the stability intervals of the 2^n and 2^{n+1} periodic solution.

It can be applied to the class of oscillators for which harmonic balance analysis yields the possibility of a period-doubling bifurcation $\lambda_1 \neq 0$ in (5.26).

Problems

1. Solve the following equation

$$\frac{\mathrm{d}r}{\mathrm{d}t} = -[(a - a_c)r + r^3],$$

$$\frac{\mathrm{d}\theta}{\mathrm{d}t} = \omega$$

for $r(0) = r_0$ and $\theta(0) = 0$. Show that at $a = a_c$ we have a supercritical Hopf bifurcation and that for $a < a_c$ the radius of the limit cycle is given by

$$r = \sqrt{|a - a_c|}.$$

2. A characteristic equation of the dynamical system linearized around the fixed point is given by

$$\lambda^4 + \alpha_3\lambda^3 + \alpha_2\lambda^2 + \alpha_1\lambda + \alpha_0 = 0,$$

where $\alpha_{0-3} > 0$. Show that a necessary conditions for a Hopf bifurcation to occur are

$$\alpha_1\alpha_2\alpha_3 - \alpha_1^2 - {\alpha_3}^2\alpha_0 = 0,$$

$$\alpha_3\alpha_2 - \alpha_1 = 0.$$

3. The bifurcation through an eigenvalue -1 (period-doubling) is generally governed by a map of the form

$$x_{n+1} = -(1 + r)x_n + ax_n^2 + bx_n^3 + \ldots$$

Show, by applying linear stability analysis, that bifurcation is subcritical for $(a^2 + b) < 0$ and supercritical for $(a^2 + b) > 0$.

4. Observe iterates of the map

$$x_{n+1} = 1 - 2x_n - \frac{1}{2}\pi(1-r)\cos\left[2\pi\left(x_n - \frac{1}{12}\right)\right] \quad \text{mod } (1).$$

Show that for $r = -0.001$ long chaotic transient behaviour is present, and observe chaotic iterates for $r = 0.001$. Intermittency observed here is of Type III and it is characterised by the existence of long-lived transient chaos below the threshold of chaotic behaviour.

5. The map introduced in Problem 4 has a fixed point at $x_{\mathrm{f}} = 1/3$. Expand the map in the neighbourhood of it into a power series and:
   (a) show that x_{f} is linearly stable for $r < 0$ and unstable for $r > 0$,
   (b) show that the period-doubling bifurcation at $r = 0$ is subcritical.
6. Consider the model for certain chemical reactions (the Brusselator). Its equations are

$$\dot{x} = a - (b+1)x + x^2 y,$$

$$\dot{y} = bx - x^2 y,$$

in which x and y are positive parameters and $x, y \geq 0$. Can this system undergo Hopf bifurcation?

6. Applications

Chaotic behaviour occurs in a great number of practical engineering and natural systems. In this chapter we briefly present several examples of chaotic behaviour in mechanical engineering, chemical reactions, electronic circuits, civil engineering problems and fluid dynamics. These examples show the variety of possible applications of chaotic and fractal dynamics in different branches of engineering. They can be considered as starting points for readers' own research in a chosen branch.

6.1 Chaos in Systems with Dry Friction

In this section, the emphasis will be on the treatment of systems described by differential equation of the form

$$\frac{\mathrm{d}^2x}{\mathrm{d}t^2} + h(x)\frac{\mathrm{d}x}{\mathrm{d}t} + f(x) = 0, \tag{6.1}$$

where f represents the restoring force and $h(x)\mathrm{d}x/\mathrm{d}t$ the damping of the system. With $T = (\mathrm{d}x/\mathrm{d}t)^2/2$ and

$$U(x) = \int^{x} f(s)\mathrm{d}s$$

we find that (6.1) yields

$$\frac{\mathrm{d}}{\mathrm{d}t}(T+U) = -h(x)\left(\frac{\mathrm{d}x}{\mathrm{d}t}\right)^2$$

If $h(x) \geq 0$, the energy $E = T+U$ decreases monotonically for free oscillations of the system, and the system tends asymptotically to the stable equilibrium position (when of course U has a minimum). However, in many vibration problems, h does not take only positive values but also negative ones. The energy of the system then does not necessarily decrease monotonically – it can also increase. In particular, there may also exist isolated periodic solutions, which we generally call a limit cycle, and this type of behaviour is termed self-excited oscillations.

To understand such oscillation phenomena we assume that the oscillations are quasiperiodic, that is that they are given by

$$x(t) = a(t)\cos[\omega t + \phi(t)],$$

where ω is constant and $a(t)$ and $\phi(t)$ are 'slowly' varying parameters. It is also assumed that the physical process is of a type for which the damping terms can be separated from the excitation term in $h\mathrm{d}x/\mathrm{d}t$. In that case, both the lost energy E_l, due to damping during a full oscillation, and the energy E_p, provided by the excitation mechanism, may be computed as functions of the amplitude a. With these assumptions one obtains

$$\int_0^T h(x(t))\left(\frac{\mathrm{d}x}{\mathrm{d}t}\right)^2 \mathrm{d}t = E_l - E_p$$

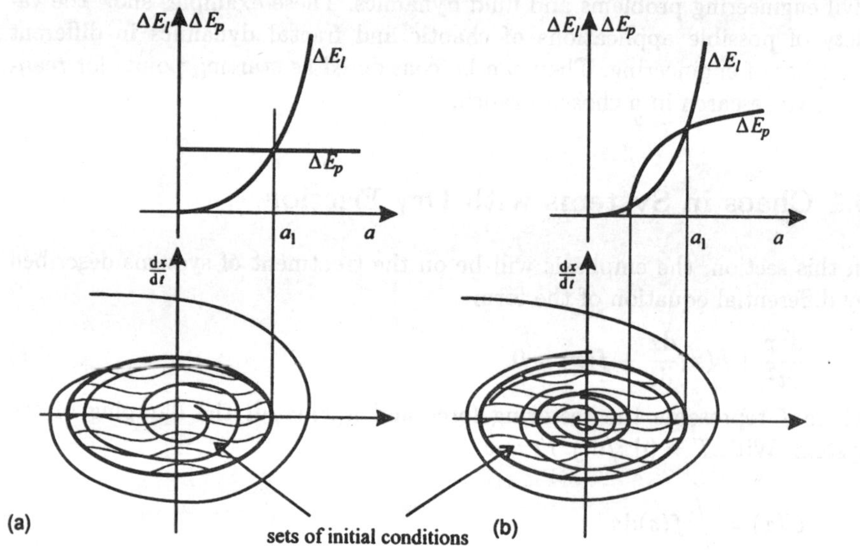

Fig. 6.1. Energy diagrams and limit cycles. (a) Stable. (b) Stable (large one) and unstable (small one)

and for a periodic solution one must have $E_l = E_p$, from which the amplitude of a limit cycle can be determined (Fig. 6.1a,b). In transient oscillation of the system, the amplitude will be increasing or decreasing depending on whether E_l or E_p is dominant. In Fig. 6.1a one observes the limit cycle that is stable, and in Fig. 6.1b there are stable (larger one) and unstable (smaller one) limit cycles.

Now consider a very simple example of a mechanical system in which the friction is negative in a certain region shown in Fig. 6.2. On a belt moving uniformly with velocity v_0 there lies a mass m fixed by the spring with stiffness k. The friction force exerted by the belt on the load is certainly a very

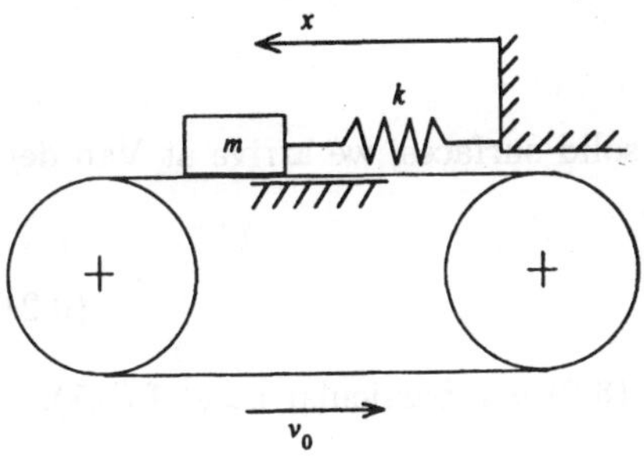

Fig. 6.2. Mechanical self-excited system

complicated function of the relative velocity of the belt and the body. If we denote the displacement of the load by x and its velocity by $\mathrm{d}x/\mathrm{d}t$, then the frictional force acting (dry friction) on the mass m, being a function of the relative velocity $v = \mathrm{d}x/\mathrm{d}t - v_0$, can be written as

$$F\left(\frac{\mathrm{d}x}{\mathrm{d}t} - v_0\right).$$

If we denote the coefficient of elasticity by k and consider that all the remaining frictional forces acting in this system (for example, the resistance of the air or the internal friction of the springs), then the equation of motion of the mass m is written as

$$m\frac{\mathrm{d}^2x}{\mathrm{d}t^2} + b\frac{\mathrm{d}x}{\mathrm{d}t} + kx + F(v) = 0,$$

where $F(v) = F(\mathrm{d}x/\mathrm{d}t - v_0)$ is a function characterising the dependence of the frictional force on the relative velocity v. Expanding the function F in a series about the value of v_0 and considering only one term of this series

$$F\left(\frac{\mathrm{d}x}{\mathrm{d}t} - v_0\right) = F(-v_0) + \frac{\mathrm{d}x}{\mathrm{d}t}\frac{\mathrm{d}F}{\mathrm{d}v_0} + \ldots$$

within the limits of this restriction, the equation of motion takes the form

$$m\frac{\mathrm{d}^2x}{\mathrm{d}t^2} + \left(b - \frac{\mathrm{d}F}{\mathrm{d}v_0}\right)\frac{\mathrm{d}x}{\mathrm{d}t} + kx = -F(-v_0).$$

The constant term occurring on the right-hand side only causes a displacement of the position of equilibrium by the amount $-F(v_0)/k$ in the direction of motion of the belt. With a transformation

$$x = -\frac{F(-v_0)}{k} + y$$

the following equation results:

$$\frac{\mathrm{d}^2y}{\mathrm{d}t^2} + \left(\beta - \frac{\mathrm{d}F}{\mathrm{d}v_0}\right)\frac{\mathrm{d}y}{\mathrm{d}t} + \omega_0^2 y = 0$$

with $\beta = k/m$, $\omega_0^2 = k/m$. So long as $[\beta - (\mathrm{d}F/\mathrm{d}v_0)] < 0$ the equilibrium position $y = 0$ is unstable. If we take

$$\frac{dF}{dv_0} = -\delta y^2 + \alpha,$$

which is characteristic of the friction of solid surfaces, we arrive at Van der Pol's equation

$$\frac{d^2y}{dt^2} + (\delta y^2 - \gamma)\frac{dy}{dt} + \omega^2 y = 0, \tag{6.2}$$

where $\gamma = \alpha - b$. Van der Pol's equation (6.2) is a particular case of (6.1).

There are a number of theorems that guarantee the existence of limit cycles for certain classes of equations [6.1], for example the Poincaré–Bendixon theorem of Chap. 2.

Chaotic behaviour can be presented only in the forced Van der Pol's equation, as according to the Poincaré–Bendixon theorem the only possible solutions of (6.2) are a fixed point or a limit cycle.

We consider the driven system

$$\frac{d^2x}{dt^2} + \mu\omega_0(x^2 - 1)\frac{dx}{dt} + \omega_0^2 x = p\cos(\Omega t)\,. \tag{6.3}$$

When a self-excited system is subjected to an additional external excitation with frequency Ω, then the system may respond with periodic oscillations with frequency Ω under certain circumstances.

It is convenient here to choose the initial point for time t in such a way that

$$\frac{d^2x}{dt^2} + \mu\omega_0(x^2 - 1)\frac{dx}{dt} + \omega_0^2 x = p_1 \sin\Omega t + p_2\cos\Omega t \tag{6.4}$$

is valid instead of (6.3). Suppose that

$$x(t) = A\sin\Omega t \tag{6.5}$$

with constant A, is a solution of (6.4). The introduction of (6.5) into (6.4) and some rearrangement yields

$$\begin{aligned} &- A\Omega^2 \sin\Omega t + \mu\omega_0 A\Omega\left[\left(\frac{A^2}{4} - 1\right)\cos\Omega t + \left(\frac{A^2}{4}\right)\cos 3\Omega t\right] \\ &+ \omega_0^2 A\sin\Omega t = p_1\sin\Omega t + p_2\cos\Omega t\,. \end{aligned} \tag{6.6}$$

This algebraic equation has no solution for non-zero p_1, p_2 and μ. However, if one compares only the terms in Ωt and thus satisfies (6.6) approximately, then one obtains

$$\begin{aligned} A(\omega_0^2 - \Omega^2) &= p_1 \\ \mu\omega_0\Omega A(A^2/4 - 1) &= p_2 \end{aligned}$$

and with $p^2 = p_1^2 + p_2^2$,

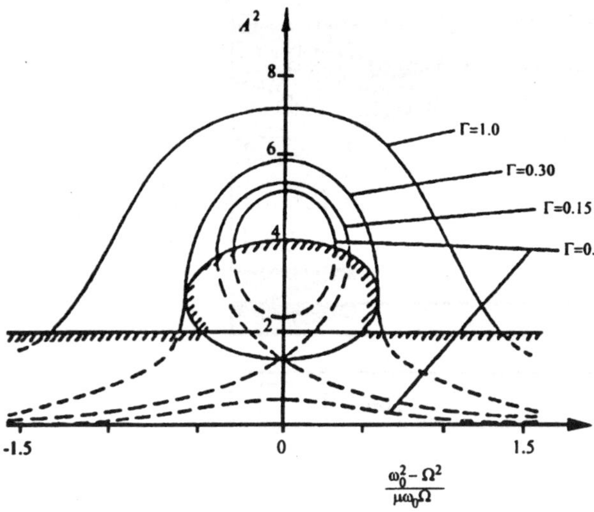

Fig. 6.3. Resonance curves of system (6.4)

$$\frac{p^2}{4\omega_0^2\mu^2\Omega^2} = \frac{A^2}{4}\left[\left(\frac{\omega_0^2-\Omega^2}{\omega_0\mu\Omega}\right)^2 + \Omega^2\left(\frac{A^2}{4}-1\right)^2\right].$$

The resonance curves A^2 versus

$$\phi = \frac{(\omega_0^2-\Omega^2)}{\mu\omega_0\Omega}$$

are shown in Fig. 6.3 for various values of the parameter $\Gamma = p^2/4\mu^2\omega_0^2\Omega^2$. For sufficiently small values of ϕ and for given Γ there are three different values of A^2. A stability investigation by means of the method described in Sect. 5.5 shows that the two smallest values of A in Fig. 6.3 correspond to the unstable solutions and the larger value to a stable oscillation. For larger ϕ only one stationary solution exists, which may be stable in some regions and unstable in others. For small Γ no stable solution exists, and for those values of system parameters for which the solution (6.5) is not stable, one can expect more complicated solutions and even chaotic behaviour.

Let us consider the dynamics of (6.3), which after the transformation $x = x_1$, $\mathrm{d}x/\mathrm{d}t = x_2$, $x_3 = \Omega t$ can be rewritten in the form

$$\begin{aligned}\frac{\mathrm{d}x_1}{\mathrm{d}t} &= x_2,\\ \frac{\mathrm{d}x_2}{\mathrm{d}t} &= -\mathrm{d}(x_1^2-1)x_2 - \omega_0^2 x_1 + p\cos x_3,\\ \frac{\mathrm{d}x_3}{\mathrm{d}t} &= \Omega. \qquad (6.7)\end{aligned}$$

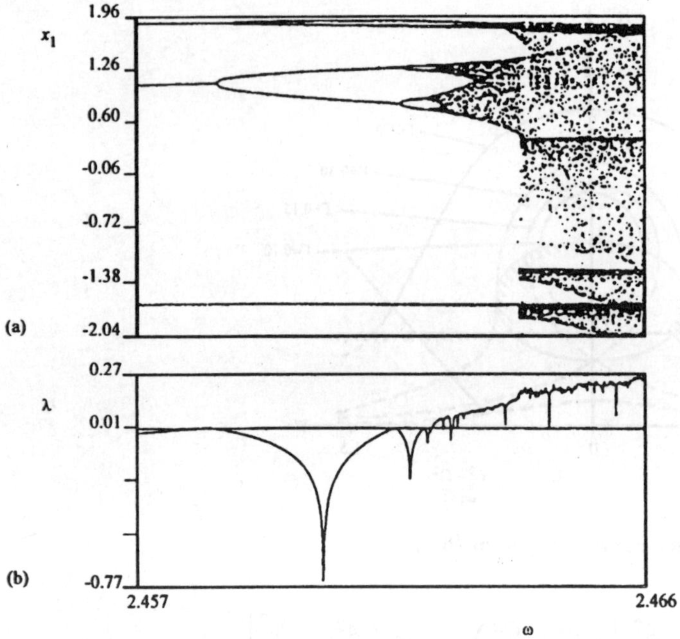

Fig. 6.4. Van der Pol's equation. (a) Bifurcation diagram. (b) Largest nonzero Lyapunov exponent versus ω

The bifurcation diagram of Fig. 6.4a shows the projections of the attractors in the Poincaré cross-section into the coordinate x_1. The plot of the largest nonzero Lyapunov exponent versus ω is shown in Fig. 6.4b. Starting with the following values of system parameters $\mu = 5$, $\omega_0 = 1$, $p = 5$ and $\omega = 2.457$ we observe a period 4 oscillation. With further increase of ω the system undergoes period-doubling bifurcations. Examples of 4×2^n, $n = 0, 1, 2,$ periodic oscillations are shown in Fig. 6.5a–c. This cascade of period-doubling bifurcations arrives at the chaotic attractor shown in Fig. 6.5d. First we observe a strange attractor that is asymmetric, but with a further small increase of ω we arrive at the strange symmetrical attractor. For more details see [6.2].

Chaotic behaviour has also been found in the generalised Van der Pol's equation

$$\frac{\mathrm{d}^2x}{\mathrm{d}t^2} - a(1-x^2)\frac{\mathrm{d}x}{\mathrm{d}t} + bx + cx^3 = p\cos\Omega t \tag{6.8}$$

for example, for $a = 0.2$, $b = 0$, $c = 1$, $p = 17$ and $\Omega = 4$ [6.6].

Now we analyse the influence of the small cubic term cx^3 on the chaotic behaviour of (6.8). We set $\mu = 5$, $\omega_0 = 1$, $p = 5$ and $\Omega = 2.466$. This problem has some practical significance as in many practical problems we

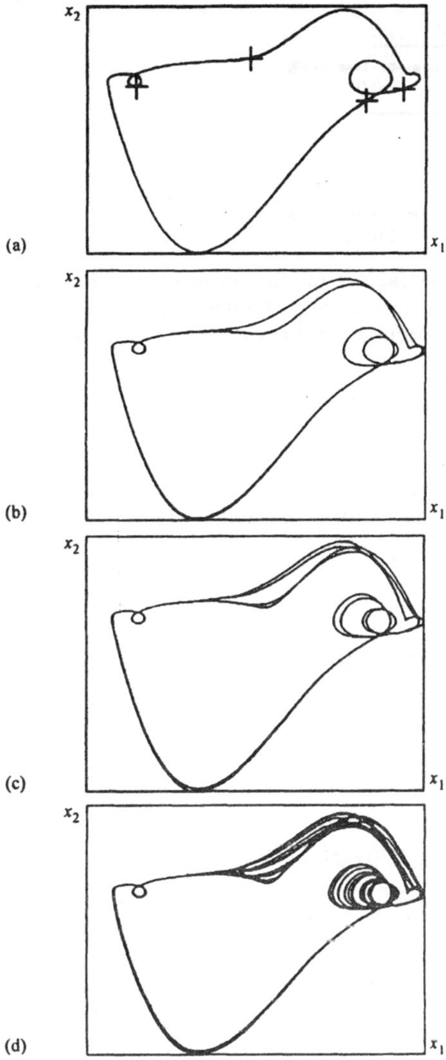

Fig. 6.5a–d. Period-doubling casacde. (**a**) $\omega = 2.457$. (**b**) $\omega = 2.460$. (**c**) $\omega = 2.462$. (**d**) $\omega = 2.463$

have to deal with a restoring force of the type $bx + cx^3$ (see Chap. 1). For a weakly nonlinear system (c small) we usually linearise this relation.

Figure 6.6 shows the behaviour of this system as a function of c. The chaotic behaviour is marked with a star and the periodic behaviour is marked with a full circle.

More details on chaotic behaviour of systems with dry friction can be found in Wiercigroch [6.3], Blazejczyk-Okolewska et al. [6.4], and Wiercigroch and de Kraker [6.5].

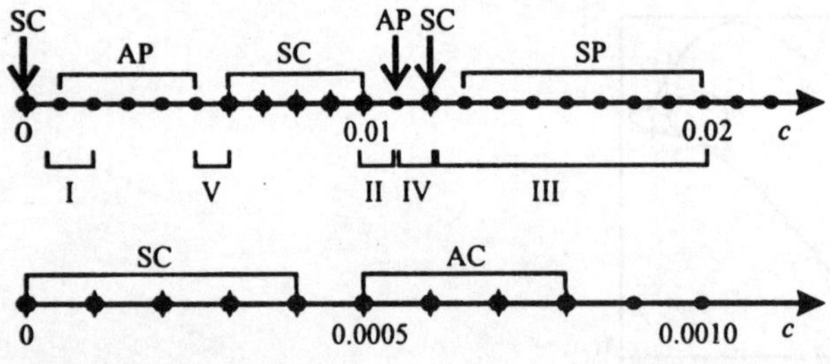

Fig. 6.6. Chaotic (•) and periodic (•) attractors on the parameter c line: AC, asymmetrical chaotic; AP, asymmetrical periodic; SP, symmetrical periodic; SC, symmetrical chaotic; I-V zones where transitions from chaotic to periodic behaviour (or vice versa) take place

6.2 Chaos in Chemical Reactions

The temporal behaviour of chemical reactions is modelled by kinetic rate equations. For the very simple reaction,

$$A \underset{}{\overset{k_f}{\rightleftharpoons}} B,$$

the time dependence of the concentrations is given by

$$\frac{dA}{dt} = -k_f A + k_r B,$$

$$\frac{dB}{dt} = k_f A - k_r B,$$

where A and B are used to denote concentrations k_f is the rate constant for the reaction $A \to B$, and k_r is the rate constant for $B \to A$. For more complicated reactions the equations become nonlinear, and its expected that under certain conditions these reactions may be chaotic.

The Belousov–Zhabotinskii (BZ) reaction is the best known example of a chemical system which exhibits both periodic and chaotic behaviour. The simplest model of it is as follows

$$A + B \overset{k_f}{\rightleftharpoons} C,$$

The reactants A and B are put into a closed container with flow rate r, and an exit port relieves the system of excess material. The rate equations,

$$\frac{dA}{dt} = -k_f AB + k_r C - r(A - A_0)$$

$$\frac{dB}{dt} = k_f AB + k_r C - r(B - B_0)$$

$$\frac{dC}{dt} = k_f AB - k_r C - rC,$$

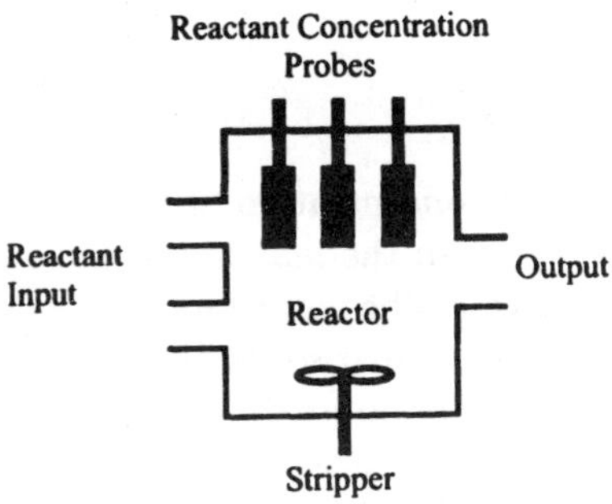

Fig. 6.7. Experimental arrangement of a chemical reaction with reactant flow. The probes monitor the reactant concentrations. (Adapted from [6.4])

exhibit nonlinear coupling between the chemical concentrations. A_0 and B_0 are the reactant concentrations at the input port ($C_0 = 0$). The experimental arrangement is shown in Fig. 6.7.

If r is zero the reaction proceeds to equilibrium, and for large r the materials are exhausted from the container before they have time to react. For intermediate r the system has both periodic and chaotic states. In this sense, the input flow rate of reactants is the control parameter analogous to the forcing amplitude of the pendulum. The temperature-dependent rate constants and initial conditions also affect the dynamical state. A phase space can be constructed for the BZ reaction that allows the periodic and chaotic behaviour to be studied. An example is shown in Fig. 6.8.

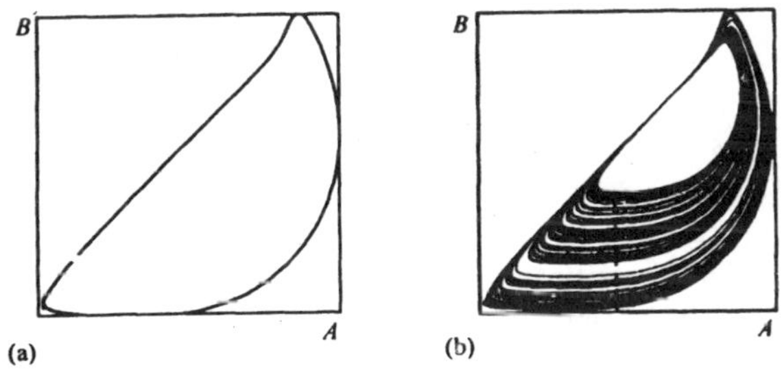

Fig. 6.8. Phase space trajectories for BZ reaction. **(a)** periodic. **(b)** chaotic behaviour

The existence of chaos in the BZ reaction suggests that similar behaviour might occur for other chemical oscillators, such as those found in a biological system. Chaotic behaviour in these systems may indicate a pathological condition, and therefore an analysis of chaotic reactions may prove useful in the study and treatment of diseases [6.8].

As our second example, we consider a model for cubic autocatalysis, developed by Gray and Scott [6.9] for a simple reaction in a closed vessel which converts the reactant P to the product C via the intermediates A, B according to the scheme

$$\begin{array}{llll} P & \to A & \text{rate} = k_0 p \\ A + 2B & \to 3B & \text{rate} = k_1 a b^2 \\ B & \to C & \text{rate} = k_2 b. \end{array} \qquad (6.9)$$

In the type of chemical system envisaged the concentration of the initial reactant P is many orders of magnitude greater than the maximum concentrations attained by the intermediates A, B, with (to be consistent) the rate of conversions from P to A being relatively slow in comparison with the other reaction rates k_1, k_2. Consequently we can, as a good approximation, regard the concentration P as being constant and equal to its initial value p_0, i.e we are making the 'pooled chemical' approximation. The differential equations governing the reaction scheme then become

$$\frac{da}{dt'} = k_0 p_0 - k_1 a b^2$$
$$\frac{db}{dt'} = k_1 a b^2 - k_2 b, \qquad (6.10)$$

where ab are the concentrations of AB respectively and t' is time. Equations (6.10), are made non-dimensional by writing

$$a = x(k_2/k_1)^{1/2}, \quad b = y(k_2/k_1)^{1/2} \quad \text{and } t' = k_2 t$$

so that (6.10) becomes

$$\frac{dx}{dt} = \mu - xy^2$$
$$\frac{dy}{dt} = xy^2 - y, \qquad (6.11)$$

where

$$\mu = (k_0 p_0/k_2)(k_1/k_2)^{1/2}$$

is a constant of order unity. Equations (6.10) were originally proposed by Sel'kov [6.10] as a simple model for oscillations in glycolysis, and their solution has been considered for all $\mu > 0$ in some detail in [6.11]. Strictly, (6.9) implies that p decays very slowly requiring μ to be proportional to $\exp(-\alpha t)$ $(\alpha > 1)$. This scheme has been studied by Merkin et al. [6.11], where a full discussion of the application of such a scheme to chemical reaction is given. Our basic assumption is to set α to be zero. If we assume μ to be oscillating periodically about μ_0 with frequency Ω, in an attempt to model the effects on the reaction due to fluctuating external conditions, we consider the system

$$\frac{dx}{dt} = \mu_0(1 + \epsilon \cos \Omega t) - xy^2$$
$$\frac{dy}{dt} = xy^2 - y. \qquad (6.12)$$

A typical set of results for $\mu = 0.95$ and $\omega = \sqrt{2}/10$ is shown in Fig. 6.9.

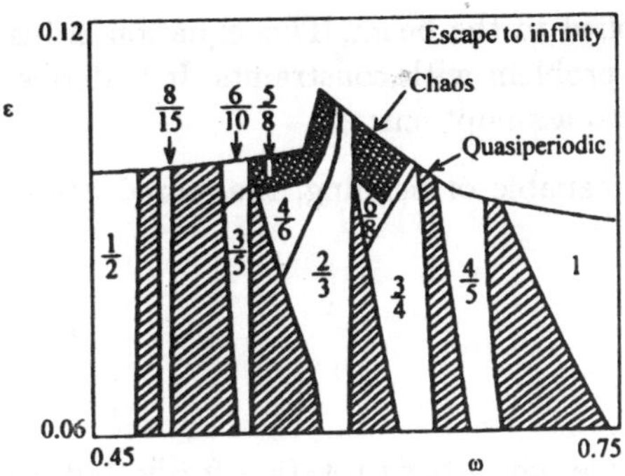

Fig. 6.9. Different types of behaviour of (6.12)

In $\epsilon - \omega$ parameter space we can distinguish between chaotic, quasiperiodic (with two incommensurate frequencies), periodic (with different m/n periods) behaviours. Additionally we can observe oscillations with increasing amplitude (escape to infinity).

6.3 Elastica and Spatial Chaos

The spatial behaviour of beams is a fundamental problem in civil engineering. The configuration of the beam, assumed planar, is most conveniently described by $u(\xi)$, the angle the beam makes with the horizontal, as a function of arc length ξ (see Fig. 6.10). Let us normalize the rod to have length π. The displacement $(x(\xi), y(\xi))$ may be calculated from the formulae

$$x(\xi) = \int_0^{\xi} \cos u(\xi')\mathrm{d}\xi', \qquad y(\xi) = \int_0^{\xi} \sin u(\xi')\mathrm{d}\xi'.$$

Equilibria of the beam are characterized by the two-point boundary problem

$$-\frac{\mathrm{d}^2 u}{\mathrm{d}\xi^2} - \lambda \sin u = 0; \quad u'(0) = u'(\pi) = 0, \tag{6.13}$$

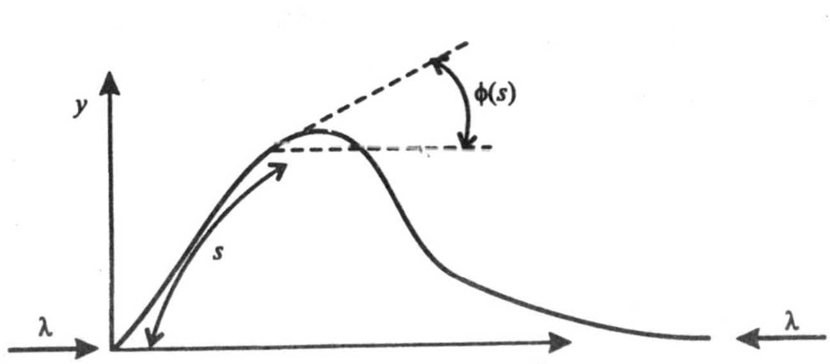

Fig. 6.10. Coordinates on the beam

where λ is the compressive force applied to the beam. This equation is just the first variation of a minimization problem with constraints. It is derived in [6.14], [6.21] under the following two assumptions:

(i) The beam is incompressible but capable of bending, the stored energy function being proportional to

$$\int_0^{\pi} \kappa^2(\xi)d\xi \tag{6.14}$$

where κ is the curvature.

(ii) The ends of the rod are hinged free, permitting rotation freely, but are constrained to lie on a line.

The last equation happened to be mathematically identical to that of a planar mathematical pendulum. There is however one important difference between the two equations because the pendulum is an initial value problem while the elastica is strictly speaking a boundary value problem. Under the assumption that our elastica is infinite we can treat (6.13) as an initial value problem

$$\frac{d^2\phi}{ds^2} + \sin\phi = 0. \tag{6.15}$$

Now we perturb this equation first by adding periodic spatial forcing (imperfection)

$$\frac{d^2\phi}{ds^2} + \sin\phi = a\sin\omega s \tag{6.16}$$

and then by adding band-limited white noise perturbation to it

$$\frac{d^2\phi}{ds^2} + \sin\phi = a\sin\omega s + A\sum_{i=1}^{N}\sin(\nu_i s + \gamma_i), \tag{6.17}$$

where ν_i and γ_i are random variables. The second component of (6.17) is an approximation of a band-limited white noise with a spectral density

$$S(\nu) = \begin{cases} \sigma(\nu_{\max} - \nu_{\min}) & \nu \in [\nu_{\min}, \nu_{\max}] \\ 0 & \nu \notin [\nu_{\min}, \nu_{\max}], \end{cases} \tag{6.18}$$

where σ is constant, $\nu_{\min}$ and $\nu_{\max}$ are band frequencies of the noise. ν_i are independent random variables with uniform distribution on the interval $[0,\ 2\pi]$, A and ν_i are given by

$$\begin{aligned} A &= \frac{\sqrt{2\sigma}}{N} \\ \nu_i &= (i-0.5)\Delta\nu + \nu_{\min} \\ \Delta\nu &= \frac{(\nu_{\max}-\nu_{\min})}{N}. \end{aligned} \tag{6.19}$$

Some of the numerical results obtained for $a = 0.01$, $0.00 \le A \le 0.01$ and three different perturbation frequency bands $0.5 \le \nu_i \le 3.5$, $0.5 \le \nu_i \le 1.5$ and $2.5 \le \nu_i \le 3.5$, $N = 300$ are shown in Fig. 6.11 which shows a spatial

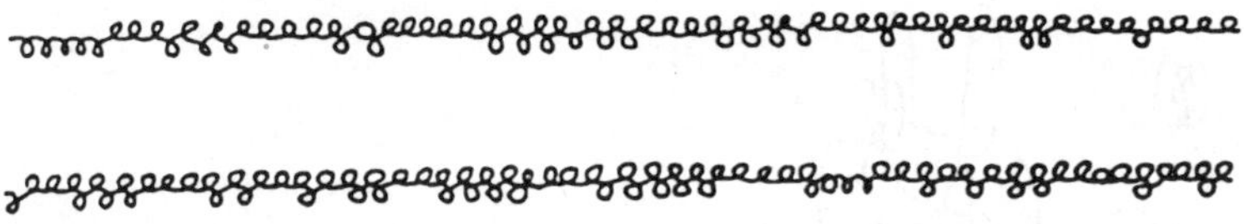

Fig. 6.11. The spatial plots for (6.15)

plot presenting the actual form which the infinite elastica should take. The results agree qualitatively with some experimental demonstrations reported in [6.13] and [6.12].

To conclude this example we consider the influence of positive damping as well as what might appear as somewhat artificial spatial forcing. This forcing arises however in a natural way in the parametric forcing of the corresponding damped pendulum problem. Thus we study the following equation of the elastica

$$\frac{\mathrm{d}^2\phi}{\mathrm{d}s^2} + \delta\frac{\phi}{\mathrm{d}s} + b\sin\phi = a\sin\omega s\sin\phi + A\sum_{i=1}^{N}\cos(\nu_i s + \gamma_i) \tag{6.20}$$

for different parameter values as well as initial conditions.

For the deterministic and noise perturbed nonlinear dynamics we plot the corresponding spatial strange attractor in the region of the strange attractor [6.15]. Figure 6.12 shows clearly the immense richness of information which these looping patterns can produce for infinitely long s.

6.4 Electronic Circuits and Chaos

In the last few years various nonlinear electronic systems which show chaotic behaviour have been constructed and described theoretically (for a great number of examples see Ogorzalek [6.16]). Certain effects blamed on noise are really examples of chaotic behaviour of a completely deterministic nature.

One such circuit is Chua's circuit (Chua [6.17], Madan [6.18], Matsumoto [6.19] and Kocarev [6.20]) which is given in Fig. 6.13. It is a third-order RLC circuit with four linear elements (two capacitors, one resistor and one inductor), and has only one nonlinear element: a piecewise linear resistor. The state equations of Chua's circuit are as follows

$$\begin{aligned} C_1\frac{\mathrm{d}v_{C_1}}{\mathrm{d}t} &= G(v_{C_2} - v_{C_1}) - f(v_{C_1}) \\ C_2\frac{\mathrm{d}v_{C_2}}{\mathrm{d}t} &= G(v_{C_1} - v_{C_2}) + i_L \\ L\frac{\mathrm{d}i_L}{\mathrm{d}t} &= v_{C_2}, \end{aligned} \tag{6.21}$$

where $G = 1/R$ and a three segment piecewise linear $v_{C_1} - i$ characteristic of nonlinear element is defined by

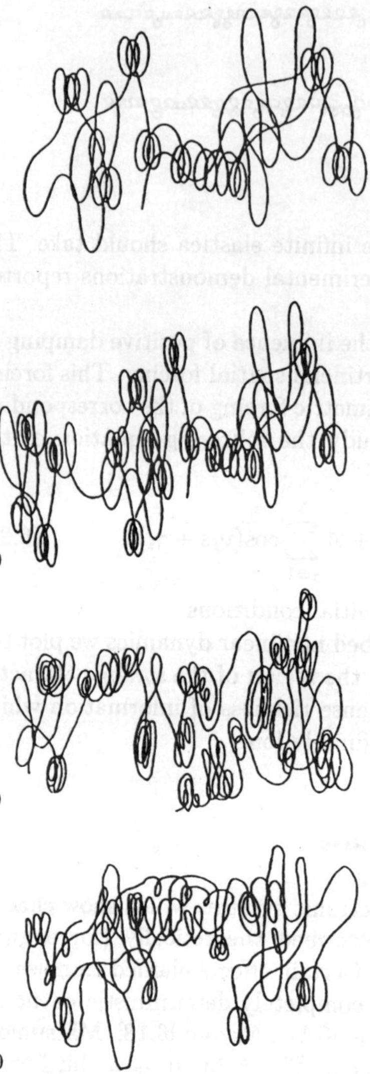

Fig. 6.12a–d. The spatial plots for (6.20): $\delta = 0.15,\ b = 1,\ a = 0.94$. (a) $\omega = 1.56,\ A = 0$. (b) $\omega = 1.58,\ A = 0$(c) $\omega = 1.56,\ A = 0.1$. (d) $\omega = 1.58,\ A = 0.15$

$$f(v_{C_1}) = m_0 v_{C_1} + \frac{1}{2}(m_1 - m_0)[|v_{C_1} + B_p| - |v_{C_1} - B_p|].$$

This relation is shown graphically in Fig. 6.14; the slopes in the inner and outer regions are m_0 and m_1 respectively; $\pm B_p$ denotes the break points. Equations (6.21) constitute an autonomous dynamical system, meaning that there is no external signal injected into the system; the system is allowed to evolve through its natural dynamics. When the resistance R, inductance L, and capacitances C_1 and C_2 in Chua's circuit are positive numbers, from an

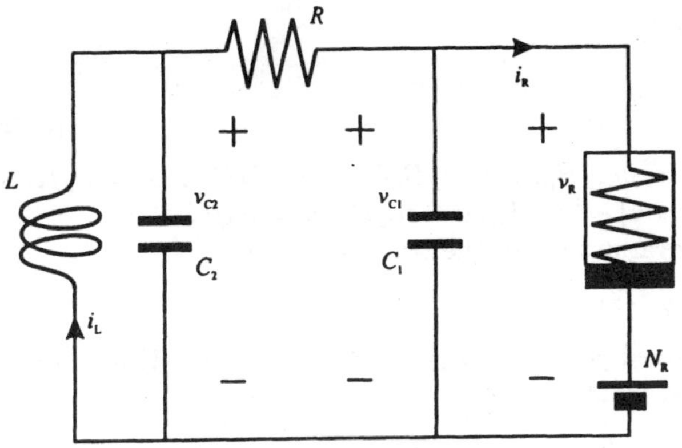

Fig. 6.13. Chua's circuit

energy-conservation point of view, the nonlinear resistor must be active for the circuit to oscillate, let alone become chaotic. In practical implementations, the active nonlinear resistor responsible for supplying power to the passive linear elements is in turn powered by a battery.

Chua's circuit is readily constructed at low cost using standard 'off-the-shelf' electronic elements and, as we shall see below exhibits a rich variety of bifurcation and chaos. This circuit is the first physical system, whose theoretical behaviour agrees with both computer simulations and experimental results, for which the presence of chaos has been proven mathematically.

Just as the classical parallel RLC resonant circuit is the simplest physical system which can model the onset of oscillations in a dynamical system, so Chua's circuit is the simplest paradigm for studying nonperiodic phenomena in nonlinear circuits. Chua's circuit is the simplest possible in the sense that chaos cannot occur in an autonomous circuit (modelled by nonlinear state equations) with fewer than three energy storage elements (capacitors and inductors) and that at least one nonlinear active element is needed for oscillation to be possible at all.

Equations (6.21) can be rewritten in the dimensionless form

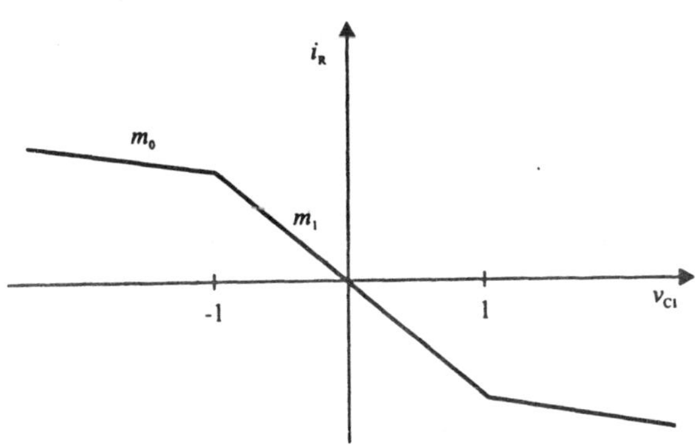

Fig. 6.14. Characteristic of nonlinear element

$$\dot{x} = \alpha[y - x - f(x)]$$
$$\dot{y} = x - y + z$$
$$\dot{z} = -\beta y, \quad (6.22)$$

where $x = v_{C_1}/B_p$, $y = v_{C_2}/B_p$, $z = i/B_pG$, $\alpha = C_2/C_1$, $\beta = C_2/G^2L$ $f(x) = bx + \frac{1}{2}(a-b)[|x+1| - |x-1|]$, $a = m_1/G$, $b = m_0/G$.

Equation (6.22) is invariant under the transformation

$$(x, y, z) \to (-x, -y, -z)$$

The origin $(0,0,0)$ is a fixed point. For $a \neq b$, $b \neq -1$, and $(a+1)(b+1) < 0$, there are two other fixed points

$$P^+ = \left(\frac{b-a}{b+1}, 0, -\frac{b-a}{b+1}\right)$$

$$P^- = \left(-\frac{b-a}{b+1}, 0, \frac{b-a}{b+1}\right).$$

The eigenvalues of the flow linearised near each fixed point are the roots of the equation

$$s^3 + s^2(\alpha c + \alpha + 1) + s(\alpha c + \beta) + \beta\alpha(c+1) = 0,$$

where c is equal to a if the fixed point is at the origin, and c is equal to b if the fixed point is at P^+ or P^-. The fixed points are stable if

$$\alpha c + \alpha + 1 > 0,$$
$$\alpha c + \beta > 0,$$
$$\alpha(c+1) > 0,$$

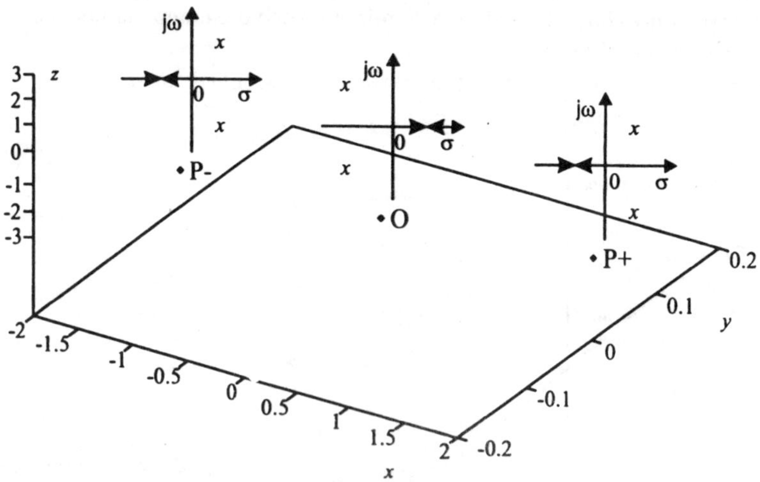

Fig. 6.15. Eigenvalue configuration of fixed points of (6.22)

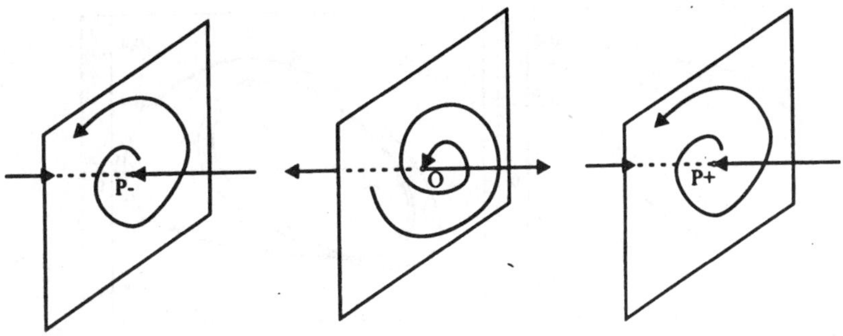

Fig. 6.16. A schematic view of the flow near fixed points

$$(\alpha c+\alpha+1)(\alpha c+\beta)>\alpha(c+1).$$

For the values of parameters in a neighbourhood of $\alpha=9$, $\beta=14.87$, $a=-8/7$ and $b=-5/7$ all fixed points are hyperbolic. The flow linearised around the origin has one positive real eigenvalue, and a complex conjugate pair of eigenvalues with negative real parts. The flow linearised around P^+ or P^- has one negative real eigenvalue and a complex conjugate pair of eigenvalues with positive real part. Eigenvalues configurations are shown in Fig. 6.15 and a typical trajectory in a small neighbourhood of each fixed point is shown in Fig. 6.16. Chua's circuit is endowed with an unusually rich repertoire of nonlinear dynamical phenomena, including all of the standard bifurcations and routes to chaos. In Fig. 6.17 we present the homoclinic trajectory of the origin which can be observed for $\alpha=11.0917459$, $\beta=14.284284$, $a=-8/7$ and $b=-5/7$.

Typical period-doubling bifurcations leading to chaos can be observed in Figs. 6.18. In this case we took $\beta=16$, $a=-8/7$ and $b=-5/7$. Fig.

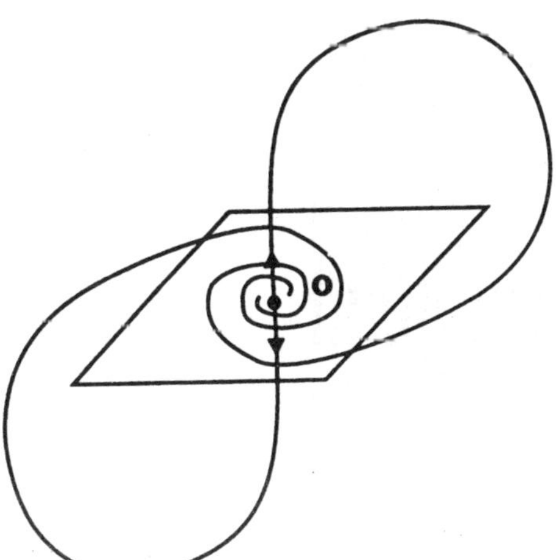

Fig. 6.17. Homoclinic trajectory of the origin in Chua's circuit

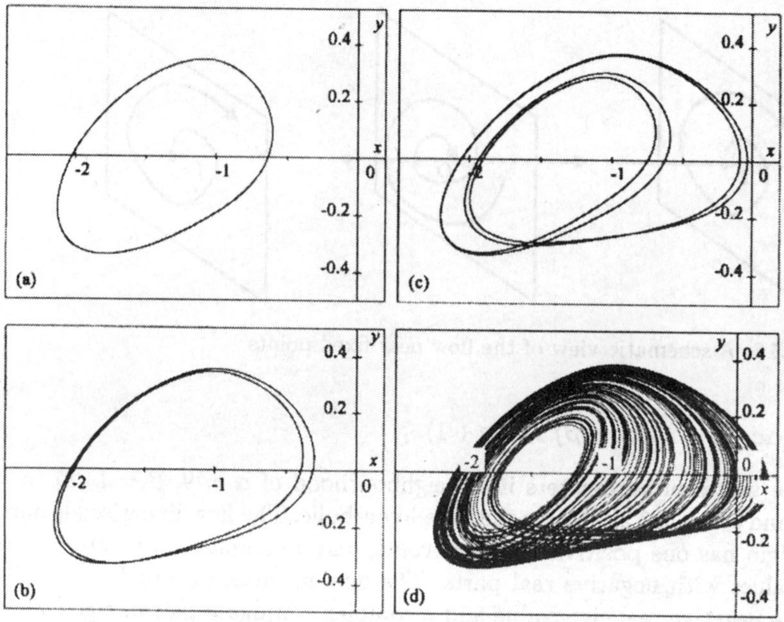

Fig. 6.18a–d. Period-doubling bifurcation in Chua's circuit. (a) Period-1. (b) Period-2. (c) Period-4. (d) Chaos

6.18a presents period-1 orbit for $\alpha = 8.8$. Period-2 ($\alpha = 8.86$), and period-4 ($\alpha = 9.12$) are shown respectively in Figs. 6.18b and 6.18c. Chaotic spiral attractor observed for $\alpha = 9.4$ is shown in Fig. 6.18d. In Fig. 6.19 we present

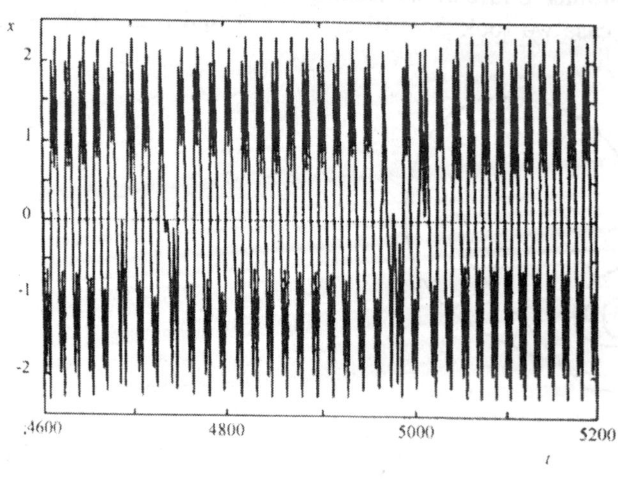

Fig. 6.19. Intermittency in Chua's circuit

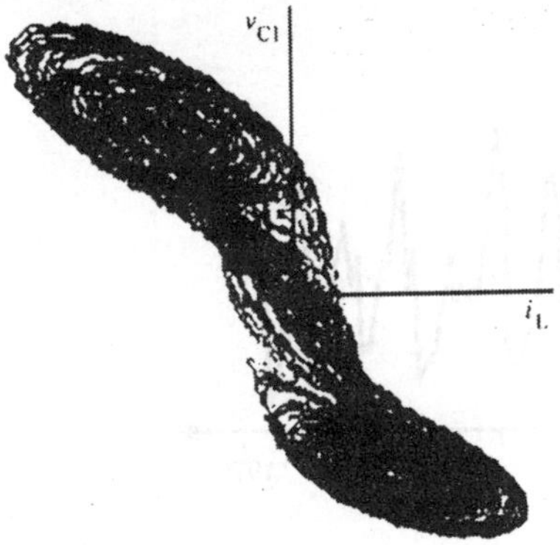

Fig. 6.20. Double scroll attractor

an example of intermittency observed for $\alpha = 4.295$, $\beta = 5$, $a = -1.27$ and $b = -0.68$.

For $\alpha = 10$, $\beta = 14.87$, $a = -1.27$ and $b = -0.68$ Chua's circuit operates on the chaotic double scroll attractor shown in Fig. 6.20.

6.5 Chaos in Model of El Nino Events

The difficulty of carrying out long-term predictions of atmospheric dynamics and the evolution of climate is a problem of obvious concern. Nowadays there is increasing awareness that deterministic chaos might provide a natural prototype for complexity of atmospheric and climatic dynamics.

In this section we describe the so-called El Nino Southern Oscillation phenomenon and show that in simple cases it can be described as a chaotic system.

El Nino may be defined as the appearance of anomalously warm water in the eastern equatorial Pacific. Associated with this is a weakening and sometimes a reversal of the trade wind field. Major El Nino Southern Oscillation events occurred in 1957, 1965, 1972 and 1982. The various events differ in detail and intensity but appear to have broadly similar overall features. El Nino has major economic consequences and possibly global climatic effects. This event can be determined by observation of some meteorological data. For example Fig. 6.21 demonstrates the variability of rainfall at Nauru Island in the West Pacific. The time series shown is characteristic, as usually very little, if any, precipitation falls in this region of the central Pacific, and only during El Nino do large amounts of precipitation occur.

Now we describe a simple but realistic model which explains all the broad qualitative features of the phenomenon. Imagine an equatorial ocean to be a box of fluid characterised by temperatures in the east and west (T_e and T_w)

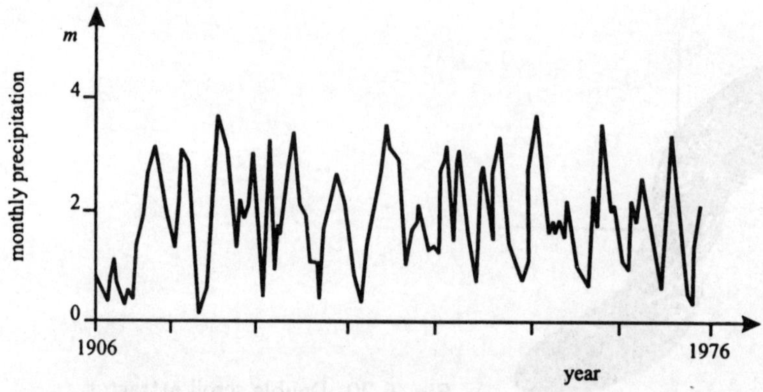

Fig. 6.21. Mean monthly precipitation at Nauru Island

and a current u as shown in Fig. 6.22. The current is driven by a surface wind which is in part generated by the temperature gradient $(T_e - T_w)\Delta x$, where Δx is the distance between points e and w. A cooler temperature in the east $(T_e < T_w)$ produces a westward surface wind across the ocean, because of the convective tendency for air to rise (sink) over warm (cool) water. Thus we write:

$$\frac{du}{dt} = \frac{B(T_e - T_w)}{2\Delta x} - C(u - u^*), \tag{6.23}$$

where B and C are constants. The terms $B(T_e - T_w)/\Delta x + Cu^*$ represent wind-produced stress, $-Cu$ represents mechanical damping and a negative value for the constant u would represent the effect of the mean tropical sur-

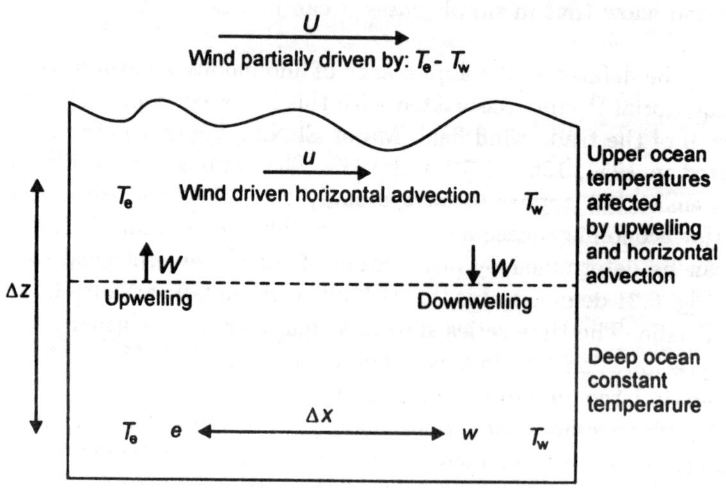

Fig. 6.22. Schema of the model

face easterlies. Variations in pressure have been neglected as they do not qualitatively affect the model behaviour.

The temperature field is advected by the current. Assuming a deep ocean of constant temperature T, the simplest finite difference approximation to the temperature equation of fluid flow is

$$\begin{aligned}\frac{\mathrm{d}T_{\mathrm{w}}}{\mathrm{d}t} &= \frac{u(T - T_{\mathrm{e}})}{2\Delta x} - A(T_{\mathrm{w}} - T^*) \\ \frac{\mathrm{d}T_{\mathrm{e}}}{\mathrm{d}t} &= \frac{u(T_{\mathrm{w}} - T)}{2\Delta x} - A(T_{\mathrm{e}} - T^*).\end{aligned} \tag{6.24}$$

The first term on each right-hand side represents horizontal advection and upwelling and the second forcing and thermal damping (A and T^* are constants). T^* is the temperature to which the ocean would relax in the absence of motion and therefore represents radiative processes and heat exchange with the atmosphere.

Equations (6.23) and (6.24) constitute the basic model in which some physical processes have been neglected. For example, small-scale structures, certain types of wave propagation, and the role of absolute temperature have not been represented.

Without loss of generality, we can set $T = 0$ thus measuring all temperatures with respect to T^*. For $u^* = 0$ it is possible to obtain some qualitative insight to the model's behaviour. Steady solutions are found by setting the left-hand sides of (6.23) and (6.24) to zero and solving the resulting equations. For

$$\frac{B}{2\Delta x^2} < \frac{AC}{T^*}$$

only one real solution exists, because the influence of the ocean temperature on the wind is so small, the state is stable and has $u = 0$ and $T_{\mathrm{e}} - T_{\mathrm{w}} = T$. If we let B increase with other parameters constant, the system will bifurcate when

$$\frac{B}{2\Delta x^2} = \frac{AC}{T^*}.$$

The existing solution becomes unstable and two new, real, stable solutions appear. All the solutions become unstable for

$$\frac{B}{2\Delta x^2} > \frac{(4A + C)C^2}{T^*(C - 2A)},$$

when the system undergoes a subcritical Hopf bifurcation. As the system is bounded, T_{w}, T_{e} and u cannot exceed certain values, we can expect chaotic behaviour.

Consequently for constants T^*, C and A all stationary solutions become unstable when B, the influence of ocean temperatures on the wind, is sufficiently large. The system can no longer stably reside anywhere and oscillate apparently randomly between the two least unstable states.

When u^* is not zero, the symmetry between T_e and T_w is broken and numerical time integration of the equations yields aperiodic behaviour remarkably evocative of El Nino (Fig. 6.23). The system stays in the neighbourhood of one unstable stationary solution for many model years before flipping to an El Nino state. El Nino states are orbits around another stationary but highly unstable, solution to (6.23) and (6.24) and the system quickly returns to its normal, less unstable state. Model El Ninos occur aperiodically, but typically with an interval of a few to several model years. Thus, with neither seasonal cycling nor stochastic forcing, the basic event cycle can plausibly be reproduced.

In our numerical experiments we have used the Vallis values [6.22], namely $A = 1$ year^{-1}, $C = 0.25$ month^{-1}, $B = 2\text{m}^2\ \text{s}^{-2}\,{}^\circ\text{C}^{-1}$, $u = -0.45\text{m}\ \text{s}^{-1}$, $\Delta x = 2500\text{km}$, $T^* = 12^\circ\text{C}$ and $T = 0^\circ\text{C}$. These values characterise the background state and may have an important impact on the dynamics.

By adding to the system annual forcing and the second stochastic forcing compatible with the 20–60 day period of forced atmospheric Kelvin modes

$$u^* = -0.45 + 0.1\cos \Omega t + \sum_{i=1}^{100} \delta_i \cos(\omega_i t + \phi_i),$$

where $\delta_i, \omega_i, \varphi_i$ are random variables such that

$$\delta_i \in (0.05, 0.3), \quad 2\pi/\omega_i \in (20,\ 60) \text{ days}$$
$$\phi_i \in (0, 2\pi), \text{ and } \ 2\pi/\Omega = 365 \text{ days}$$

one obtains a more realistic model of El Nino event. It should be noted here that the effects of the oscillations of the atmosphere, associated with propagating and reflecting Kelvin waves, are important El Nino dynamics, as has been suggested by recent observational evidence [6.23].

Figure 6.23 shows the variation of u and ΔT over a period of 100 years after 1000 years of integration. Roughly, periodic large anomalies are very

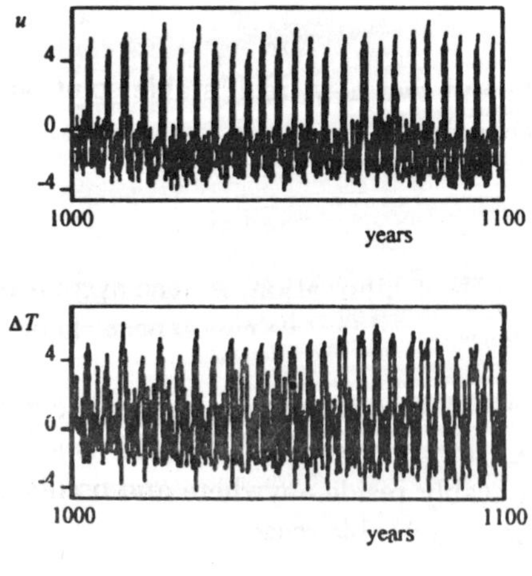

Fig. 6.23. Variation of u and ΔT over a period of 100 years as a result of integrating (6.23) and (6.24) for a 1000 values of parameters as specified in text; $\Delta T = T_e - T_w$

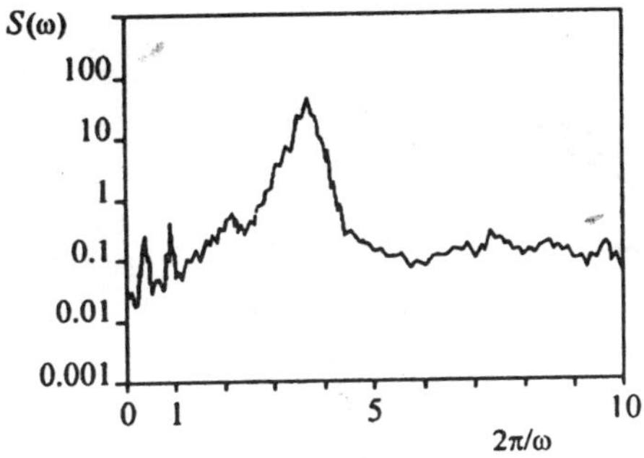

Fig. 6.24. Spectral density distributions for t over the 100 years of Fig. 6.23

prominent in these plots. Gross features are summarised in the spectrum (Fig. 6.24), which shows a strong noisy peak corresponding to a mean period of 3.4–5 years. All these features are known from meteorological observations.

Some support for the description of the real El Nino dynamics as noisy periodicity rather than chaos may be adduced by an analysis of the example of rainfall data given in Fig. 6.21. Briefly, any dissipative system such as this ocean–atmosphere system, has the property that its phase space volume contracts as the system approaches its asymptotic state (attractor). If we consider the attractor as a set of points embedded in the phase space, then by a well-known procedure described in Sect. 2.8 we can assign to it a number called its correlation dimension. The dimension turns out to be a lower bound to the number of independent variables necessary to describe the attractor. If this dimension is non-integer (as in the case of chaotic systems) then the bound is the next higher integer.

Following Grassberger-Procaccia we constructed an m-component vector from the time series $p(t)$ shown in Fig. 6.21 as

$$P_i(t_i) = p_1(t_1 + \tau), p_2(t_1 + 2\tau), \ldots, p_m(t_1 + m\tau),$$

where τ is an appropriate time delay, and calculated the correlation integral defined for N vectors distributed in an m-dimensional embedded space as a function of distance r:

$$C(r, m) = \lim_{N\to\infty} N^{-2} \sum_{i=1}^{N} \sum_{j=1}^{N} U(r - |P_i - P_j|),$$

where U is the unit step function. If the number of points is large enough, as assumed above, this distribution will obey a power-law scaling with r for small r : $C(r) \sim r^{d_{\rm corr}}$, where $d_{\rm corr}$ is the correlation dimension.

The correlation dimension can be used to quantify the presence of chaos in our system. In the case of chaos, as we increase the embedding dimension m, the correlation dimension is seen to converge to a fixed value. If the system is random the correlation dimension is seen to increase monotonically with m. Fig. 6.25 displays the results of a calculation from the time series of rainfall at Nauru shown in Fig. 6.21. There is no evidence of a low-dimensional

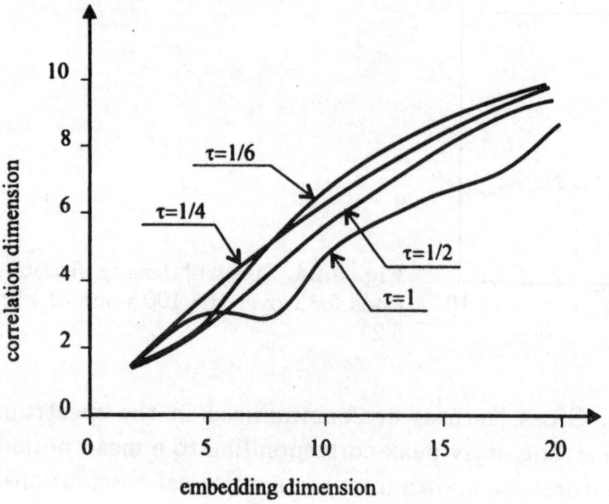

Fig. 6.25. Correlation dimension ν versus embedding dimension m for a time series of monthly mean percipitation of Nauru (see Fig. 6.21)

attractor such as is characteristic of chaotic phenomena and the suggestion is that ENSO events cannot be adequately modelled by a low-dimensional dynamical system.

The example described in this section shows the situation that is typical in the modelling of real physical systems: Chaos can be observed in the consideration of a simple low-dimensional model while adding some additional forcings or couplings which are characteristic of the phenomena, we obtain a response which can be interpreted as noisy periodicity.

7. Controlling Chaos

As was shown in previous chapter chaos occurs widely in engineering and natural systems; historically it has usually been regarded as a nuisance and designed out if possible. It has been noted only as irregular or unpredictable behaviour, and often attributed to random external influences. More recently there have been examples of the potential usefulness of chaotic behaviour. It is to the potential usefulness of chaotic behaviour that we turn our attention in this chapter.

We first review a number of methods by which undesirable chaotic behaviour may be controlled or eliminated. More speculatively, we indicate ways in which the existence of chaotic behaviour may be directly beneficial or otherwise usefully exploited. In the next section we describe methods which allow the synchronisation of two chaotic systems. The possible application of chaos for secure communication is discussed in the last section.

7.1 Controlling Methods

We can divide approaches to controlling chaos into two broad categories; firstly those in which the actual trajectory in the phase space of the system is monitored, and some feedback process is employed to maintain the trajectory in the desired mode, and secondly nonfeedback methods in which some other property or knowledge of the system is used to modify or exploit the chaotic behaviour.

7.1.1 Control Through Feedback

Ott, Grebogi and Yorke [7.1] and Romeiras [7.2] have, in an important series of papers, proposed and developed the method by which chaos can always be suppressed by shadowing one of the infinitely many unstable periodic orbits (or perhaps steady states) embedded in the chaotic attractor. The basic assumptions of this method are as follow:

(1) The dynamics of the system can be described by an n-dimensional map of the form

$$\xi_{n+1} = f(\xi_n, p). \tag{7.1}$$

In the case of continuous-time systems tis map can be constructed e.g., by introducing the Poincaré map.

(2) p is some accessible system parameter which can be changed in some small neighbourhood of its nominal value p^*.
(3) For the value of p^* there is a periodic orbit within the attractor around which we would like to stabilize the system.
(4) The position of this orbit changes smoothly with p for small variations of p.

Let ξ_F be a chosen fixed point of the map f of the system existing for the parameter value p^*. In the close neighbourhood of this fixed point with good accuracy we can assume that the dynamics is linear and can be expressed approximately by

$$\xi_{n+1} - \xi_{\mathrm{F}} = M(\xi_{\mathrm{n}} - \xi_{\mathrm{F}}). \tag{7.2}$$

The elements of the matrix M can be calculated using the measured chaotic time series and analysing its behaviour in the neighbourhood of the fixed point. Further the eigenvalues e_{s} , e_{u} and eigenvectors v_{s}, v_{u} of this matrix can be found and they determine the stable and unstable manifolds in the neighbourhood of the fixed point.

Denoting by f_{s}, f_{u} the contravariant eigenvectors ($f_{\mathrm{s}}v_{\mathrm{s}} = f_{\mathrm{u}}v_{\mathrm{u}} = 1$), $f_{\mathrm{s}}v_{\mathrm{u}} = f_{\mathrm{u}}v_{\mathrm{s}} = 0$) we can find the linear approximation valid for small $|p_n - p^*|$

$$\xi_{n+1} = p_n g + [e_{\mathrm{u}} v_{\mathrm{u}} f_{\mathrm{u}} + e_{\mathrm{s}} v_{\mathrm{s}} f_{\mathrm{s}}][\xi_n - p_n g],$$

where

$$g = \frac{\delta \xi_f(p)}{\delta p}|_{p+p^*}.$$

As ξ_{n+1} should fall on the stable manifold of ξ_{F}, we choose p_n such that $f_{\mathrm{u}}\zeta_{n+1} = 0$;

$$p_n = \frac{e_{\mathrm{u}} \xi_n f_{\mathrm{u}}}{(e_{\mathrm{u}} - 1) g f_{\mathrm{u}}}. \tag{7.3}$$

The idea of the Ott–Grebogi–Yorke algorithm is schematically described in Fig. 7.1.

The main properties of this method are as follows:

(1) This is a feedback method;
(2) Any accessible system parameter can be used as a control parameter;
(3) Noise can destabilize the controlled orbit resulting in occasional chaotic bursts;
(4) Before settling into the desired periodic orbit the trajectory exhibits a long chaotic transient.

As an example of the application of this method consider the control of chaos in Chua's circuit (described in Sect. 6.4) operating on the double-scroll

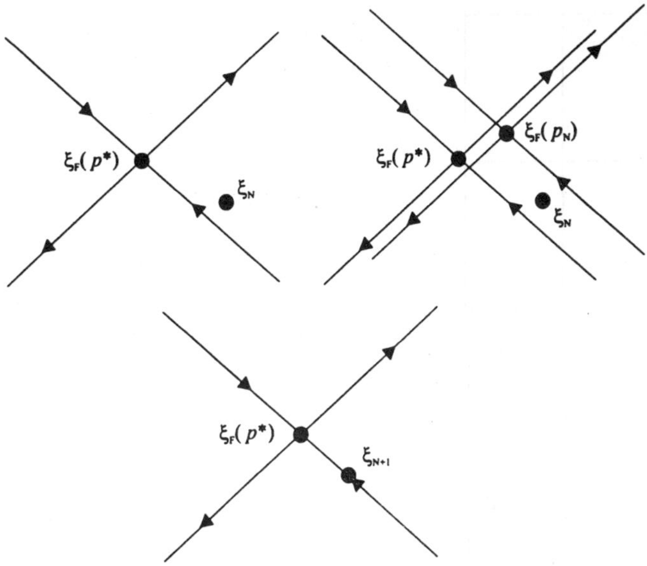

Fig. 7.1. Basis idea of Ott–Grebogi–Yorke method

attractor. The block diagram of the implemented system [7.3] is shown in Fig. 7.2.

Figure 7.3 shows the stabilisation of period 1 and period 2 unstable periodic orbits (presented in black). Before the control is achieved the trajectories exhibit chaotic transients (shown in grey).

The OGY approach has stimulated a good deal of research activity, both theoretical and experimental. The efficiency of the technique was demonstrated by Ditto et al. [7.4] in a periodically forced system, converting its chaotic behaviour into period 1 and period 2 orbits, and the application of the method to stabilise higher periodic orbits in a chaotic diode resonator was demonstrated by Hunt [7.5]. Another interesting application of the method is the generation of a desired aperiodic orbit in Mehta and Henderson [7.6] and controlling transient chaos in Tel [7.7]. Related work [7.8] used time delay techniques to control chaos.

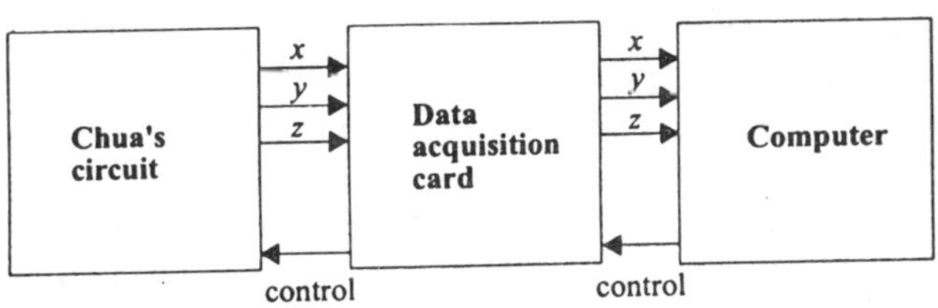

Fig. 7.2. Practical implementation of the Ott–Grebogi–Yorke method

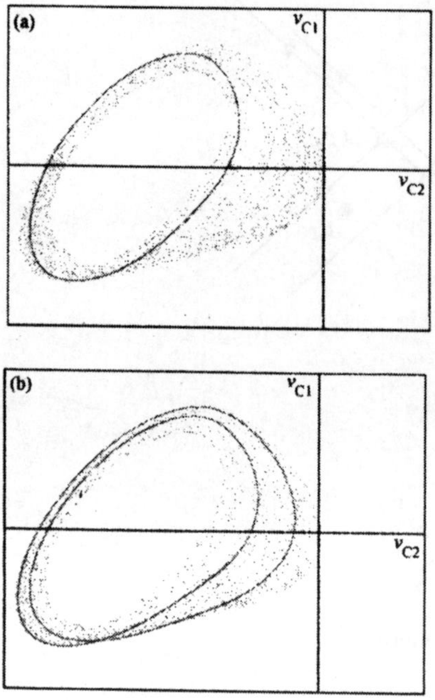

Fig. 7.3. Control of orbits. (a) Period 1. (b) Period 2

Though the OGY theory has been proposed in the context of low-dimensional dynamical systems, and most of the experimental or observational investigations have been concerned with clearly low-order mechanical or electrical contexts, the interesting experiments by Singer et al. [7.9] demonstrate its potential for fluid (and perhaps fluid–solid) mechanical phenomena. The experiments succeeded in achieving regular laminar flow in previously unstable thermal convection loops by use of a thermostat-type feedback.

Generally, the experimental application of the OGY method requires a continous computer analysis of the state of the system. The changes of the parameters, however, are discrete in time since the method deals with the Poincaré map. This leads to some serious limitations. The method can stabilise only those periodic orbits whose maximum Lyapunov exponent is small compared to the reciprocal of the time interval between parameter changes. Since the corrections of the parameter are rare and small, the fluctuation noise leads to occasional bursts of the system into the region far from the desired periodic orbit especially in the presence of noise.

A different approach to feedback control which allows us to avoid the above mentioned problems is the method of a time-continuous control proposed by Pyragas [7.10]. This method is based on the construction of a special form of a time-continuous perturbation, which does not change the form of the desired unstable periodic orbit, but under certain conditions can stabilise it. Two feedback controlling loops, shown in Fig. 7.4, have been proposed. A combination of feedback and periodic external force is used in the first

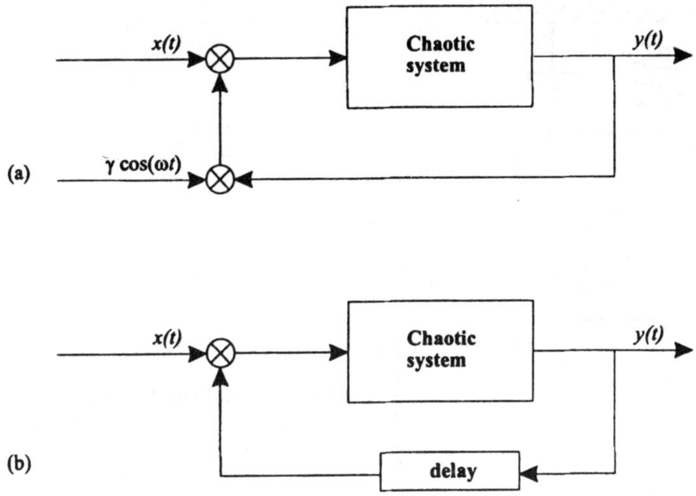

Fig. 7.4a,b. Feedback loops for continuous control of chaos

method (Fig. 7.4a). The second method (Fig. 7.4b) does not require any external source of energy and it is based on a self-controlling delayed feedback. If the period of external force $T = 2\pi/\omega$ or a time delay τ is equal to the one of the unstable periodic orbits embedded in the chaotic attractor, it is possible to find a constant K which allows stabilisation of the unstable periodic orbits.

7.1.2 Control by System Design

In this section we explore the idea of modifying or removing chaotic behaviour by appropriate system design. It is clear that, to a certain extent, chaos may be "designed out" of a system by appropriate modification of parameters, perhaps corresponding to modification of mass or inertia of moving parts. Equally clearly, there exist strict limits beyond which such modifications cannot go without seriously affecting the efficiency of the system itself.

More usefully, it may be possible to effect a control by joining the chaotic system with some other (small) system. The idea of this method is similar to that of the so-called dynamical vibration absorber. A dynamical vibration absorber is a one-degree-of-freedom system, usually mass on a spring (sometimes viscous damping is also added), which is connected to the main system as shown in Fig. 7.5. The additional degree of freedom introduced shifts resonance zones, and in some cases can eliminate oscillations of the main mass. Although such a dynamical absorber can change the overall dynamics substantially, it must be usually only physically small in comparison with the main system, and does not require an increase of excitation force. It can be easily added to the existing system without major changes of design

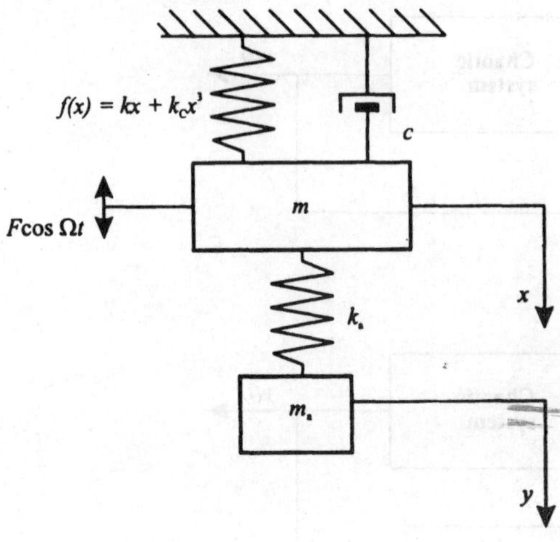

Fig. 7.5. Schematic diagram of the main system and the dynamical absorber

or construction. This contrasts with devices based on feedback control, which can be large and costly.

To explain the role of dynamical absorbers in controlling chaotic behaviour let us consider the Duffing oscillator, coupled with an additional linear system:

$$\frac{\mathrm{d}^2x}{\mathrm{d}t^2} + a\frac{\mathrm{d}x}{\mathrm{d}t} + bx + cx^3 + d(x-y) = B_0 + B_1\cos\Omega t\,, \tag{7.4a}$$

$$\frac{\mathrm{d}^2x}{\mathrm{d}t^2} + e(y-x) = 0\,, \tag{7.4b}$$

where a, b, c, d, e, B and Ω are constant. Here d and e are the characteristic parameters for the absorber, and we take e as a control parameter. The parameters of (7.4) are related to those of Fig. 7.5 in the following way: $a = c/m\Omega$, $b = k/m\Omega^2$, $c = k_c/m\Omega^2$, $d = k_a/m\Omega^2$, $e = k_a/m_a\Omega^2$, $B_0 = F_0/m\Omega$ and $B_1 = F_1/m\Omega$. It should be noted here that parameters d and e are related to each other through the absorber stiffness k_a. For simplicity in the rest of this section we assume that d is constant and consider e as control parameter, i.e. we take constant stiffness k_a and allow the absorber mass, m_a, to vary.

It is well-known that the uncoupled equation (7.4a) (i.e. without the dynamical absorber) shows chaotic behaviour for certain parameter regions (Ueda [7.11], Kapitaniak [7.12]). As has been mentioned in Sect. 5.4, in many cases the route to chaos proceeds via a sequence of period-doubling bifurcations, and in such cases our method provides an easy way of switching between chaotic and periodic behaviour.

To analyse the system with the absorber (present $d, e \neq 0$), we first assume that all parameters of (7.4), excluding the forcing frequency Ω, are constant, and estimate the Ω-domain where chaos exists. The application of the harmonic balance method enables us to determine the stability domain of appropriate $2\pi/\Omega = T$ periodic solutions, i.e.

$$\begin{aligned} x &= C_0 + C_1 \cos(\Omega t + \phi) \\ y &= D_0 + D_1 \cos(\Omega t + \gamma) \end{aligned} \tag{7.5}$$

and $2T$ periodic solutions:

$$\begin{aligned} x &= A_0 + A_{1/2} \cos[\frac{\Omega}{2} + \rho] + A_1 \cos \Omega t \\ y &= E_0 + E_{1/2} \cos[\frac{\Omega}{2} + \beta] + E_1 \cos \Omega t, \end{aligned} \tag{7.6}$$

where C_0, C_1, D_0, D_1, A_0, $A_{1/2}$, A_1, E_0, $E_{1/2}$, E_1, ϕ , γ, ρ and β are constants which are determined by substituting (7.5) or (7.6) into (7.4). Approximate boundaries of stability as functions of forcing frequency Ω for each solution can be estimated by adding small perturbations dx and dy to x and y, and considering an appropriate Hill's equation. The whole procedure is similar to the one described in Sect. 5.5, so we omit details here. Knowing the period-doubling bifurcation values Ω_1' and Ω_2' at which we have bifurcation from $T \to 2T$ periodic solutions, and Ω_1'' and Ω_2'' at which we have bifurcation from $2T \to 4T$ periodic solutions, we can obtain approximate values for the accumulation points Ω_1^∞ and Ω_2^∞ in the way described in Sect.5.5 and the interval $[\Omega_1^\infty, \Omega_2^\infty]$ can be considered as an approximation of the frequency domain for which chaos exists.

The above procedure can be easily performed using any symbolic algebra system and by following it for different values of e, we are able to obtain a map of behaviour of (7.4) as a function of two parameters: the frequency Ω and the dynamical absorber control parameter e, as shown for example in Fig. 7.5 (solid lines). The other parameters of (7.5) have been fixed at the values $a = 0.077$, $b = 0$, $c = 1.0$, $B_0 = 0.045$ and $B_1 = 0.16$. This plot is in good agreement with numerically obtained behaviour domains as shown in Fig. 7.5 (broken lines). Numerical results have been obtained using a fourth order Runge–Kutta method with a time step $\pi/200$, and to determine chaotic behaviour the Lyapunov exponents were calculated using the algorithm of Wolf et al. [7.13].

From Fig. 7.6 it is clear that, for fixed Ω, we can obtain different types of periodic behaviour by making slight changes in e. As an example, consider a system with $\Omega = 0.98$. For $e < 0.09$, the system is chaotic, but by changing e from 0.01 to 0.16 it is possible to obtain easily T, $2T$, $4T$, $8T$ periodic orbits.

Theoretically orbits of higher periods are also possible, but their narrow range of existence makes them difficult to find either experimentally or numerically. What is of vital significance is that the values of the parameter $e \in [0.01, 0.16]$ can be obtained with an absorber mass m_a approximately 100 times smaller than the main mass (Fig. 7.4). To show the effectiveness of our method in real experimental conditions we have considered the effect of quasiperiodic noise given by (6.18 and 6.19) on the system (7.4). Considering

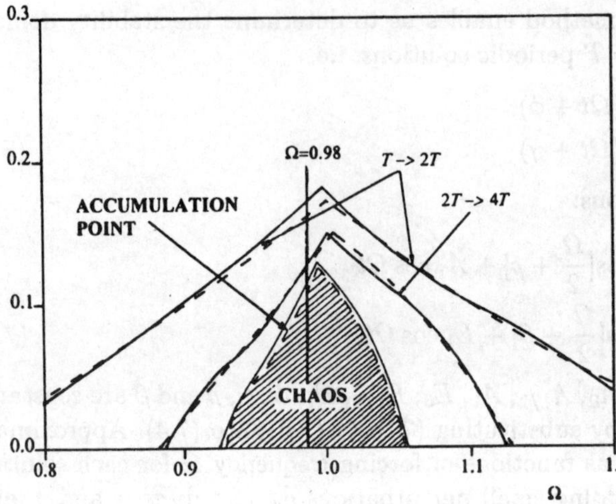

Fig. 7.6. Behaviour of (7.4) for different values of e and Ω: $a = 0.077$, $b = 0$, $c = 1.0$, $B_0 = 0.045$ and $B_1 = 0.16$; analytical approximation (*solid line*), numerical simulation (*broken line*)

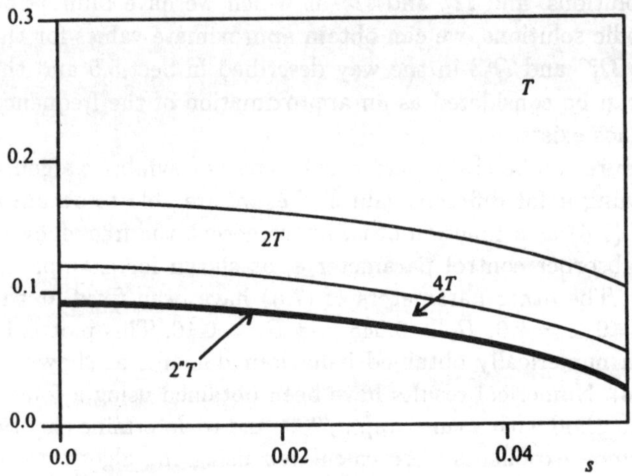

Fig. 7.7. Effect of noise on the behaviour of (7.4): $\Omega = 0.98$, other parameter values as in Fig. 7.6

the perturbed system

$$\frac{d^2x}{dt^2} + a\frac{dx}{dt} + bx + cx^3 + d(x-y)$$

$$= B_0 + B_1 \cos\Omega t + \sum_{i=1}^{N} A_i \cos(\nu_i t + \phi_i),$$

$$\frac{d^2x}{dt^2} + e(y-x) = 0 \qquad (7.7)$$

we have found the interesting property that the presence of noise reduces the magnitude of e necessary to obtain an appropriate periodic solution. This property is summarized in Fig. 7.7, where we compare the behaviour of the system (7.4) for different noise intensity.

A similar controlling effect can be obtained by varying the absorber stiffness, i.e. by simultaneous changes of parameters d and e.

7.1.3 Selection of Controlling Method

We have described several methods by which chaotic behaviour in a mechanical system can be modified, displaced in parameter space or removed. The OGY method is extremely general, relying only on the universal property of chaotic attractors that they have embedded within them infinitely many unstable periodic orbits (or even static equilibria). On the other hand, the method requires us to follow the trajectory and employ a feedback control system which must be highly flexible and responsive; such a system in the mechanical configuration can be large and expensive. It has the additional disadvantage that small amounts of noise may cause occasional large departures from the desired operating trajectory.

The nonfeedback approach is inevitably much less flexible, and requires more prior knowledge of equations of motion. On the other hand, to apply such a method, we do not have to follow the trajectory. The control procedures can be applied at any time and we can switch from one periodic orbit to another without returning to the chaotic behaviour, although after each switch transient chaos can be observed. The lifetime of this transient chaos strongly depends on initial conditions. Moreover, in a nonfeedback method we do not have to wait until the trajectory is close to an appropriate unstable orbit; in some cases this time can be quite long. The dynamic approach can be very useful in mechanical systems, where feedback controllers are often very large (sometimes larger than the control system). In contrast, a dynamical absorber having a mass of order 1 percent of that of the control system is able, as we have shown in the example of Sect. 7.1.2, to convert chaotic to periodic behaviour over a substantial region of parameter space. Indeed the simplicity by which chaotic behaviour can be changed in this way, and the possibility of easy access to different periodic orbits, may actually motivate the search for and exploitation of chaotic behaviour in practical engineering systems. This prompts us to pose a final question – how can we exploit chaos in practical systems? The OGY method, at least in theory, gives access to the wide range of possible behaviour encompassed by the unstable periodic (and other) orbits embedded in a chaotic attractor. Moreover, the sensitivity of the chaotic regime to both initial conditions and parameter values means that the desired effects can be produced by fine tuning. Thus, we may actually wish to design chaos into a system, in order to exploit this adaptability. Nonfeedback methods can, in principle, give us advice on the design, whether we wish to design chaos out or in. Additionally, they enable us to choose regions of the design parameter space or operating parameter space within which chaos will occur and be acceptable.

An example of practical use might be the minimisation of metal fatigue by switching from a necessary strictly periodic operation of the fully loaded conditions, which repeats stresses applied at the same places to a noisy periodicity (rather like a healthy heartbeat) under idling conditions.

Full description of chaos controlling procedures can be found in Kapitaniak [7.14], Chen and Dong [7.15] and Schuster [7.16].

7.2 Synchronisation of Chaos

The essential property of a chaotic trajectory is that it is not asymptotically stable. Closely correlated initial conditions have trajectories which quickly become uncorrelated. Despite this obvious disadvantage, it has been established that synchronisation of two chaotic systems is possible [7.17].

7.2.1 Pecora and Carroll's Approach

The basic synchronisation procedure [7.17] can be described as follows. Suppose that the n-dimensional dynamical system

$$\frac{\mathrm{d}u}{\mathrm{d}t} = h(u)$$

can be divided into two subsystems

$$\frac{\mathrm{d}x}{\mathrm{d}t} = f(x,y)$$
$$\frac{\mathrm{d}y}{\mathrm{d}t} = g(x,y), \tag{7.8}$$

where

$$u = (x,y)^T\,,$$
$$x = (u_1, ..., u_m)^T\,,$$
$$f = (h_1(u), ..., h_m(u))^T\,,$$
$$y = (u_{m+1}, ..., u_n)^T\,,$$
$$g = (h_{m+1}(u), ..., h_n(u))^T\,.$$

Let us create a new subsystem z identical to the y subsystem and augment (7.8) with this new system, giving

$$\frac{dx}{dt} = f(x, y)$$
$$\frac{dy}{dt} = g(x, y) \tag{7.9}$$
$$\frac{dz}{dt} = g(x, z),$$

the first two equations of (7.9) are called a *driving subsystem* and the third one *response subsystem.*

The Lyapunov exponents of the response subsystem for a particular input $x(t)$ are called *conditional Lyapunov exponents.* Let $y(t)$ be a chaotic trajectory with initial condition $y(0)$, and $z(t)$ be a trajectory started at a different initial point $z(0)$. It has been shown that the necessary and sufficient condition for

$$|z(t) - y(t)| \to 0, \tag{7.10}$$

that is for two subsystems to be synchronised, is that all of the conditional Lyapunov exponents are negative.

We can describe this procedure using an example of Chua's circuit introduced in Sect. 6.4. Its dimensionless equation (6.21) can be decomposed in three different ways described as follows:

(a) x-drive configuration where the state equation becomes

* driving subsystem

$$\frac{dx}{dt} = \alpha(y - x - f(x))$$
$$\frac{dy}{dt} = x - y + z$$
$$\frac{dz}{dt} = -\beta y,$$

* response subsystem

$$\frac{dy'}{dt} = x - y' + z'$$
$$\frac{dz'}{dt} = -\beta y';$$

(b) y-drive configuration with the state equations

* driving subsystem

$$\frac{dy}{dt} = x - y + z$$
$$\frac{dx}{dt} = \alpha(y - x - f(x))$$
$$\frac{dz}{dt} = -\beta y,$$

* response subsystem

$$\frac{dz'}{dt} = \alpha(y - x' - f(x')$$

$$\frac{dz'}{dt} = -\beta y';$$

(c) z-drive configuration where the state equation is as follows

* driving subsystem

$$\frac{dy}{dt} = -\beta y$$

$$\frac{dx}{dt} = \alpha(y - x - f(x))$$

$$\frac{dy}{dt} = x - y + z,$$

* response subsystem

$$\frac{dz'}{dt} = \alpha(y' - x' - f(x')$$

$$\frac{dy'}{dt} = x' - y' + z.$$

It can be shown that for $\alpha = 10$, $\beta = 14.87$, $a = -1.27$, $b = -0.68$ driving and response subsystems can be synchronised only in the x- and y-drive configurations, since the conditional Lyapunov exponents are $[\lambda_1^c = -0.05, \lambda_2^c = -0.05]$, $[\lambda_1^c = -2.55, \lambda_2^c = 0]$, $[\lambda_1^c = -5.42, \lambda_2^c = 1.23]$ respectively for x, y and z configurations [7.18].

7.2.2 Synchronisation by Continuous Control

Now let us describe a slightly simpler method which allows synchronisation using the continuous chaos control method described in Sect. 7.1.1.

To synchronise two chaotic systems A and B whose dynamics are described respectively by

$$\frac{dx}{dt} = f(x)$$

$$\frac{dy}{dt} = f(y), \tag{7.11}$$

where $x, y \in \mathcal{R}^n$, we use the strategy which is schematically illustrated in Fig. 7.8. We assume that some state variables of both systems A and B can be measured. Let us say that we can measure the signal $x_i(t)$ from the system A and the signal $y_i(t)$ from B where $i = 1, 2, ..., n$. Chaotic systems A and B are coupled unidirectionally in such a way that the difference $D(t)$ between the signals $x_i(t)$ and $y_i(t)$ is used as a control signal

$$K(x_i(t) - y_i(t)) = KD(t), \tag{7.12}$$

which is applied to one of the chaotic systems (A in Fig. 7.8) as a negative feedback. The parameter $K > 0$ is an experimentally adjustable weight of the

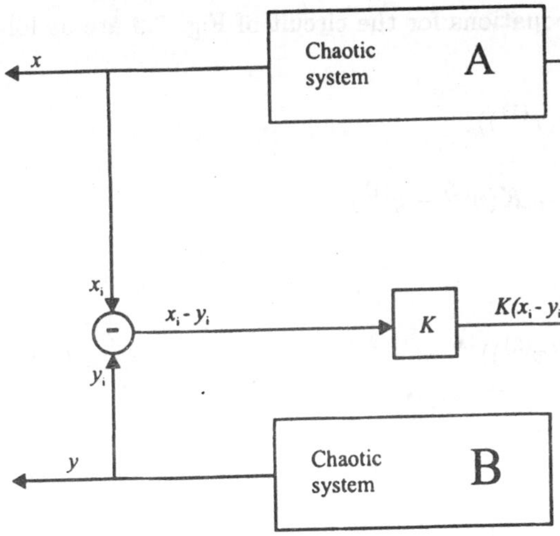

Fig. 7.8. Scheme of our method of synchronisation of two chaotic systems. Some dynamical variables of the two systems ($x_i(t)$ and $y_i(t)$) are measured and chaotic systems A and B are coupled via negative feedback

perturbation and we discuss its selection later. An experimental realisation of such a feedback presents no difficulties for many practical systems. An important feature of the perturbation signal (7.12) is that it does not change the solution of (7.11). When the synchronisation is achieved $F(t)$ becomes zero so the chaotic systems A and B become practically uncoupled.

We illustrate our synchronisation procedure using two identical unidirectionally coupled Chua's circuits, as shown in Fig. 7.9.

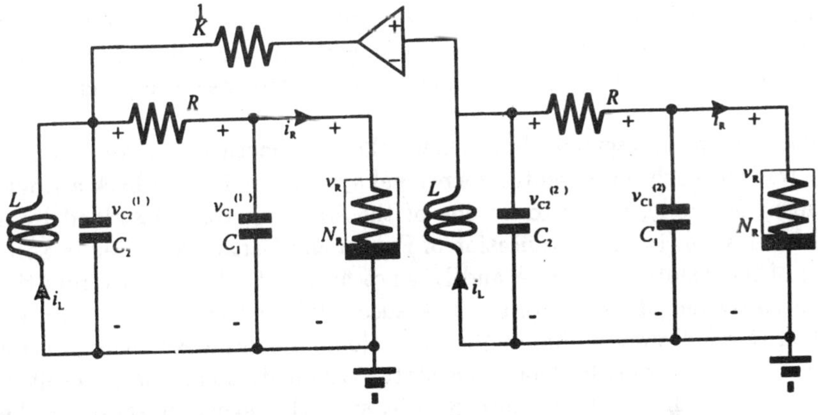

Fig. 7.9. Two identical unidirectionally coupled Chua's circuits

The nondimensional state equations for the circuit of Fig. 7.9 are as follows:

$$\begin{aligned}
\frac{dx^{(1)}}{dt} &= \alpha(y^{(1)} - x^{(1)} - f(x^{(1)})) \\
\frac{dy^{(1)}}{dt} &= x^{(1)} - y^{(1)} + z^{(1)} + K(y^{(2)} - y^{(1)}) \\
\frac{dz^{(1)}}{dt} &= -\beta y^{(1)} \\
\frac{dx^{(2)}}{dt} &= \alpha(y^{(2)} - x^{(2)} - f(x^{(2)})) \\
\frac{dy^{(2)}}{dt} &= x^{(2)} - y^{(2)} + z^{(2)} \\
\frac{dz^{(2)}}{dt} &= -\beta y^{(2)},
\end{aligned} \tag{7.13}$$

where superscripts (1) and (2) denote state variables of respectively A and B Chua's circuits. We have assumed that for $K = 0$ both circuits operate on the double-scroll attractors of Fig. 6.21. If $[x^{(1)}(0), y^{(1)}(0), z^{(1)}(0)]^T$ is slightly different to $[x^{(2)}(0), y^{(2)}(0), z^{(2)}(0)]^T$, the state trajectories of the two circuits diverge exponentially from each other.

To achieve synchronisation we add the perturbation signal

$$K(y^{(2)} - y^{(1)}) \tag{7.14}$$

to the second equation of (7.13).

In Fig. 7.10 we showed two $x^{(1)}$ versus $x^{(2)}$ plots of system (7.13). Observe from Fig. 7.10a that for $K = 1.2$ this plot is a straight line which indicates that synchronisation of the two Chua's circuits is achieved. For smaller values of K, for example $K = 1.1$, the structure of the $x^{(1)}$ versus $x^{(2)}$ plot is more complicated, and the two circuits in Fig. 7.10b are not synchronised.

Our numerical simulations are in good agreement with the experimental results, as can be seen in Fig. 7.11 where we presented the corresponding v_{C_1} versus v_{C_2} [7.19].

The example presented above shows that our method is a very convenient way to synchronise multi dimensional systems by feeding back a single variable. Generally, it has to be one of the state variables described by a drive subsystem (in the classification of Pecora and Carroll described in Sect. 7.2.1) of the chaotic systems A and B, since feeding back variables from the driven subsystem does not result in a successful continuous chaos control method [7.20]. However, due to the limitations of continuous chaos control methods [7.20], feeding back only one state variable does not always result in a successful synchronisation. One can easily show that synchronisation can be achieved only if the number of positive Lyapunov exponents of the 'composite' coupled system is equal to the number of positive Lyapunov exponents of the component systems.

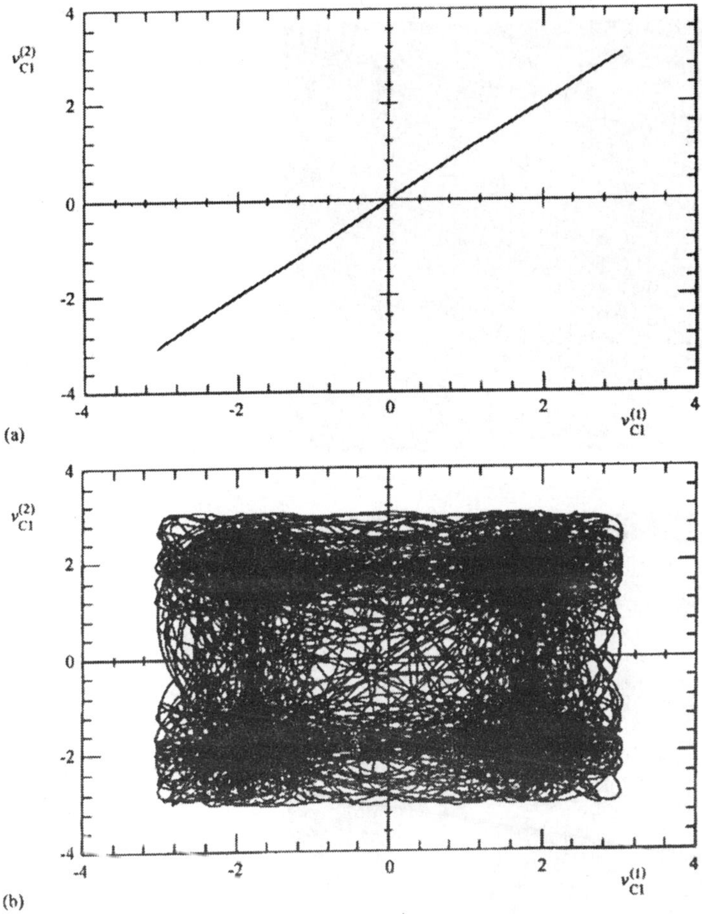

Fig. 7.10. Numerical $x^{(1)}$ versus $x^{(2)}$ plots of unidirectionally coupled Chua's circuits. (a) $K = 1.2$. (b) $K = 1.1$

In our example only one positive Lyapunov exponent in the spectrum of (7.13) is allowed in order to achieve synchronisation. Knowing the equations of chaotic systems A and B we can easily check the above condition by direct computation of the Lyapunov exponents. If the Lyapunov exponents condition is fulfilled, then the coupled system (7.13) will evolve on the same manifold on which both chaotic systems evolve and this is why synchronisation can be obtained. When it is not fulfilled, the coupled system will evolve on a higher-dimensional manifold on which a chaotic attractor with more than one positive Lyapunov exponents exists and synchronisation cannot be obtained.

The region of synchronisation in the K-parameter space can be enlarged if we simultaneously feed back more state variables of the chaotic systems

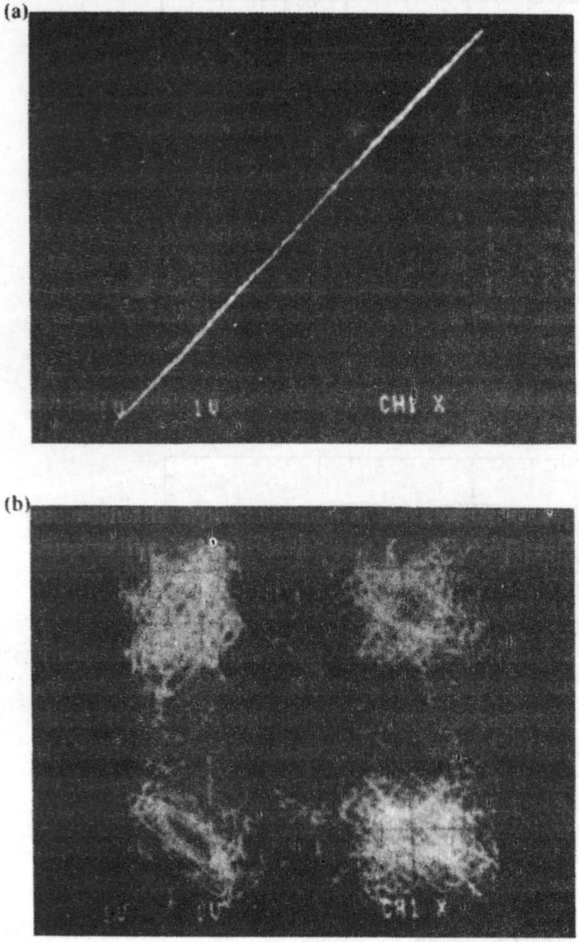

Fig. 7.11. Experimental $v_{C1}^{(1)}$ versus $v_{C2}^{(2)}$ plots of unidirectionally coupled Chua's circuits. **(a)** $K = 1.2$. **(b)** $K = 1.1$

A and B. Such an example is shown in Fig. 7.12. In this figure we showed experimental and numerical plots of system trajectories for $K = 1.1$ when, besides the perturbation (7.14) added to the second equation of (7.13), an additional

$$K(x^{(2)} - x^{(1)})$$

perturbation is added to the first equation of (7.13). It can be seen that by feeding back two state variables $x^{(1)}$ or $v_{C1}^{(1)}$ and $x^{(2)}$ or $v_{C2}^{(2)}$ synchronisation is achieved. In comparison with the previous situation when only one state variable is fed back, we observe that in the current situation the two Chua's circuits are synchronised with smaller $K < 0.9$.

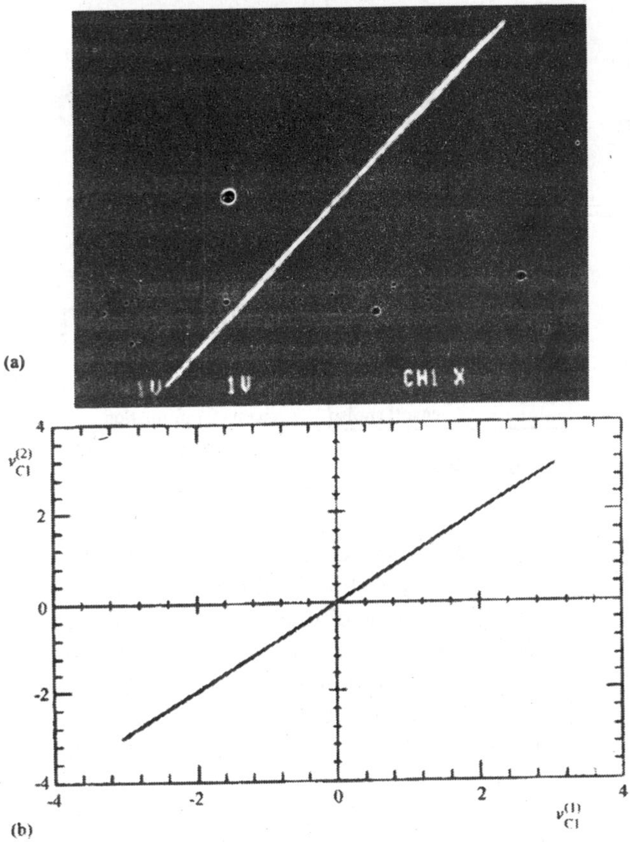

Fig. 7.12. $v_{C1}^{(1)}$ versus $v_{C2}^{(2)}$ plots of unidirectionally coupled Chua's circuits, both $v_{C1}^{(2)}$ and $v_{C2}^{(2)}$ are fed back. (a) Experimental. (b) Numerical

7.3 Secure Communication

Chaos synchronisation allows applications of chaotic systems to mask the information signal $I(t)$ by adding it to a larger chaotic signal $n(t)$ and transmitting the superposition of the two signals. Information can be recovered after the comparison of the received signal $I(t)+n(t)$ with the original chaotic noise $n(t)$. In this procedure chaotic signals in the transmitter and receiver systems must be synchronised. As this way of sending information is difficult to unmask, it is called *secure communication*. The main idea of secure communication is sketched in Fig. 7.13 (for more general description see Cuomo and Oppenheim [7.21], Kocarev et al. [7.22], and Halle et al. [7.23]). In this section we describe a simple method of transmission of digital signals proposed in [7.24]. Two chaotic systems A and B are synchronised via negative feedback. The dynamical variable $x(t)$ of the first system is transmitted as a

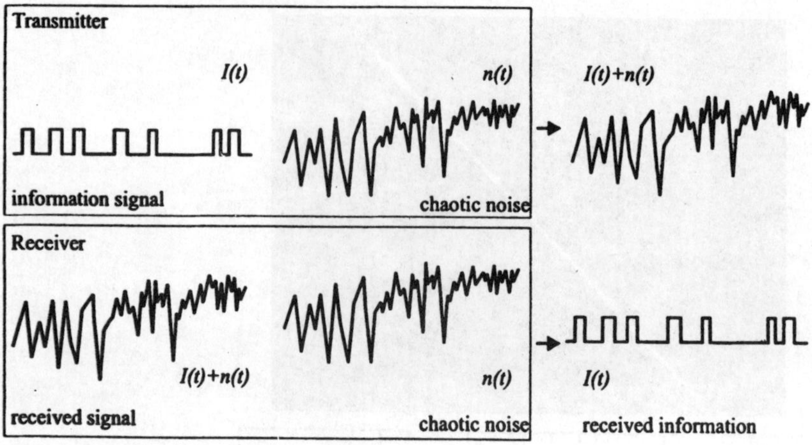

Fig. 7.13. Idea of secure communication

masking signal to which an information sequence $\delta x(t)$ is added. The comparison of the received signal $x(t) + \delta x(t)$ with a chaotic signal of system B – $y(t)$ allows information recovery. The scheme of this method is shown in Fig. 7.14. As it is known synchronisation regime is characterized by a straight line in the $x_i(t) - y_i(t)$ plot. Consider the information signal which is transmitted in the form shown in Fig. 7.15. When the function $\delta x_i(t)$ has the first jump to the positive value $h << \max(x_i(t))$ (1 in a binary sequence), i.e. the first

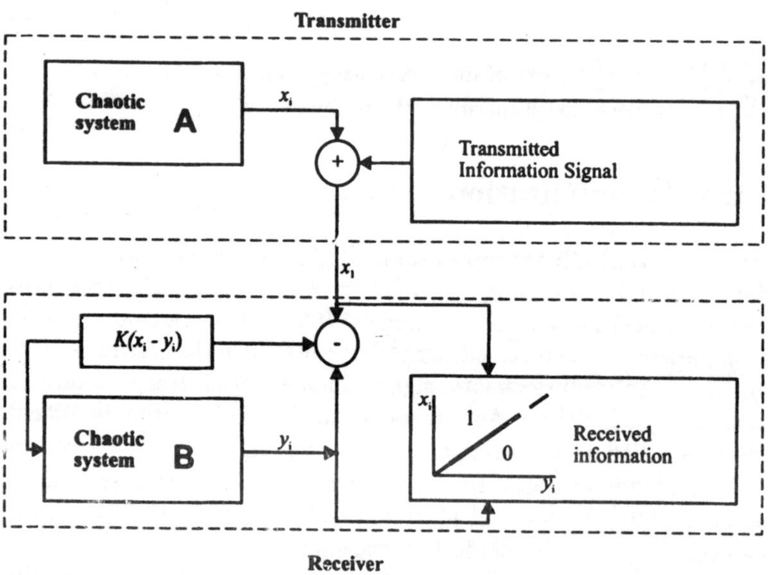

Fig. 7.14. Scheme of the method of transmitting binary sequence

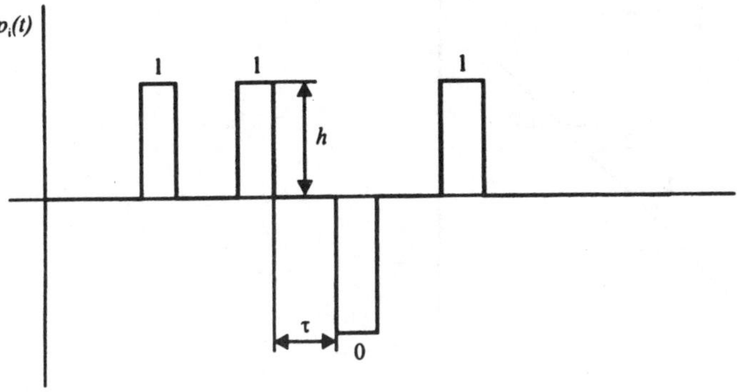

Fig. 7.15. Information signal $\delta x_i(t)$

bit of information is transmitted, the trajectory leaves synchronisation line in $x_i(t) - y_i(t)$ plot and goes into a half plane $x_i(t) > y_i(t)$ (Fig. 7.16). With this event we can associate symbol 1 in a recovered sequence. Next the value of $\delta x_i(t)$ becomes zero, the system trajectory approaches the neighbourhood of synchronisation line and after synchronisation time $\tau_s < \tau$ both systems are synchronised again and we can transmit another bit of information. When the 0 symbol is transmitted $\delta x_i(t) = -h < 0$ and in the recovering procedure we associate 0 with a departure of the trajectory from synchronisation line into the domain $x_i(t) > y_i(t)$ (Fig. 7.16). Each transmitted binary symbol can be sent after a break of duration τ in which the synchronisation of our systems is ensured. τ has to be evaluated experimentally. As was shown in [7.19] synchronisation time τ depends on coupling stiffness K and for large K this can be relatively small.

In our numerical investigations we again used two identical Chua's circuits, described by (7.13) as chaotic systems. Through this system we transmitted the binary sequence of 1010 bits with $h = \max(\delta y) = 0.5$, $\tau = 0.02$, $K_2 = 260.0$ and the transmitted signal was recovered in $y - v$ plot without errors. The procedure of recovering information is sketched in Fig. 7.17. There are several reasons to claim that the signal transmission system constructed using the chaos synchronisation concept is secure. Firstly, one could claim

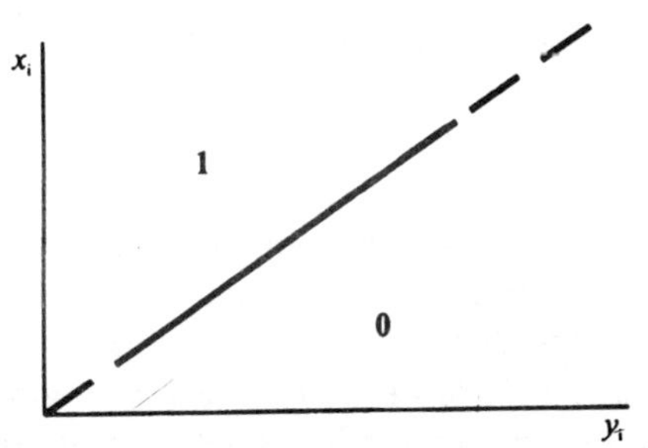

Fig. 7.16. $x_i(t) - y_i(t)$ plot used for recovering information sequence

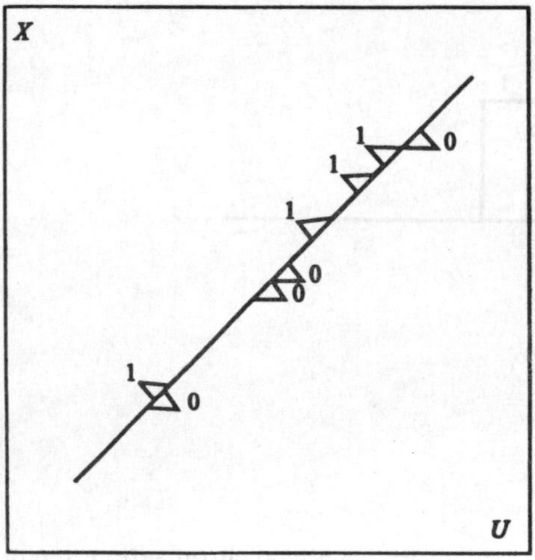

Fig. 7.17. Experimental recovery of information sequence

that the information signal is buried within the chaotic carrier and thus is not distinguishable for an observer. Even for the sender it would be hard if not impossible to decode/extract the hidden signal as the transmitted signal is chaotic, broad-band with continuous spectra, having no resemblance to the message signal and at first sight it looks noise-like. So it is quite obvious that from such a "non-expert" point of view such transmission has a substantial measure of security. As a second important security factor, one has to consider the possibilities of decoding the signal – as it has been shown already to reproduce the information signal we must know the parameters of the coder/decoder circuit with sufficient precision (if not exactly). Thus the parameters of the coding/decoding device serve as a secret key. It is obvious that in reality there exists an infinite number of parameter sets which result in chaotic outputs generated by the system. It is truly impossible to explore all these parameter sets to find the right one enabling decoding; such search for 'a needle in the haystack' would take too long.

Recently, Perez and Cerdeira [7.25] have published some results showing that for most primitive methods of signal transmission on a chaotic carrier (direct parameter modulation or masking), they were able to extract the information signal using partial reconstruction techniques using just the transmitted signal without any knowledge about the coder/decoder.

We have to stress here that this example was chosen for the sake of simplicity – our prime interest here was to demonstrate that the monotonic synchronisation principle works also in practical applications. We do not claim that any new scheme for increased security has been proposed. We believe however that having more sophisticated coding schemes (modulation) and/or more complicated chaotic coders/decoders (eg. hyperchaotic or cascaded chaotic circuits), which produce signals to be transmitted with virtually no correlation with the information signal, monotonic synchronisation can offer several advantages over the classical synchronisation such as short settling time (which has been demonstrated by examples), robustness to parameter mismatch and channel noise and imperfections.

7.4 Estimation of the Largest Lyapunov Exponent Using Chaos Synchronisation

Most real engineering systems are discontinuous. For example, in considering mechanical systems with dry friction, as described in Sect. 6.1 one has to include a nondifferential function of the signum type. Since the mathematical algorithms for calculating Lyapunov exponents described in Sect. 2.5 require differentiable equations of motion, we cannot use them in such systems.

In this section we propose a method of estimation of the largest Lyapunov exponent of discontinuous systems based on properties of chaos synchronisation, namely the fact that two chaotic systems coupled unidirectionally can synchronise when the coupling strength is larger than the largest Lyapunov exponent of a single system.

Let us consider a dynamical system which consist of two identical n-dimensional subsystems coupled with unidirectional feedback characterized by amplifying coefficient k. The equations describing such a system can be written as:

$$\begin{aligned} \dot{x}_1 &= f(x_1, a) \\ \dot{x}_2 &= f(x_2, a) + k(x_1 - x_2), \end{aligned} \tag{7.15}$$

where $x_1, x_2, k \in \mathbf{R}^n$ or, after substituting $x = x_1 - x_2$:

$$\begin{aligned} \dot{x}_1 &= f(x_1, a) \\ \dot{x} &= f(x, x_1, a, k), \end{aligned} \tag{7.16}$$

where x represents the state difference between coupled subsystems.

The synchronisation value of the coupling parameter k_s, is a boundary value, beyond which only full synchronisation of the two subsystems described by (7.15) takes place. Such synchronisation exists only when the state difference x has value zero. The analytical approach described below tries to estimate an approximate value of the coefficient k_s.

Let $D(x_1)$ and $D(x_2)$ denote the divergence of the separated systems when no coupling occurs, i.e. $k = 0$. Because the subsystems are identical we can write the equality:

$$D(x_1) = D(x_2) = \sum_{i=1}^{n} \lambda_i(x_1) = \sum_{i=1}^{n} \lambda_i(x_2) = g(a), \tag{7.17}$$

where λ_i denotes a Lyapunov exponent.

We now introduce the divergence D of the coupled subsystem (7.16). This is clearly given by:

$$D = \sum_{i=1}^{2n} \lambda_i(x_1) = D(x_1) + D(x_2) - \sum_{i=1}^{n} k_i, \tag{7.18}$$

or

$$D = \sum_{i=1}^{2n} \lambda_i(x_1) = D(x_1) + D(x), \tag{7.19}$$

where $D(x)$ is a contribution arising from the property of *state difference*.

Unidirectional coupling in equations (7.15) means that the divergence $D(x_1)$ does not change after changes of the k parameter value. Then we conclude that half of the Lyapunov exponents (related to $D(x_1)$) preserve their constant values even when the coupling parameter changes. These values correspond to the values of the Lyapunov exponents of the separated individual subsystems. During such changes the divergence $D(x)$ decreases when the value of the coefficient k increases, and in consequence the value of the sum of Lyapunov exponents related to *state difference* also decreases. As an example we show in Fig. 7.18 a graph of Lyapunov exponents of two coupled Lorenz systems versus the value of the parameter k. Those relating to $D(x_1)$ are represented by three horizontal lines, while the decreasing lines relate to $D(x)$.

The dependence of $D(x)$ on the value of the parameter k results from a comparison of (7.18) and (7.19), from which we get:

$$D(x) = \sum_{i=1}^{n} \lambda_i(x) = D(x_2) - \sum_{i=1}^{n} k_i. \tag{7.20}$$

Using (7.17) in (7.20) we obtain:

$$D(x) = \sum_{i=1}^{n} \lambda_i(x_i) - \sum_{i=1}^{n} k_i, \tag{7.21}$$

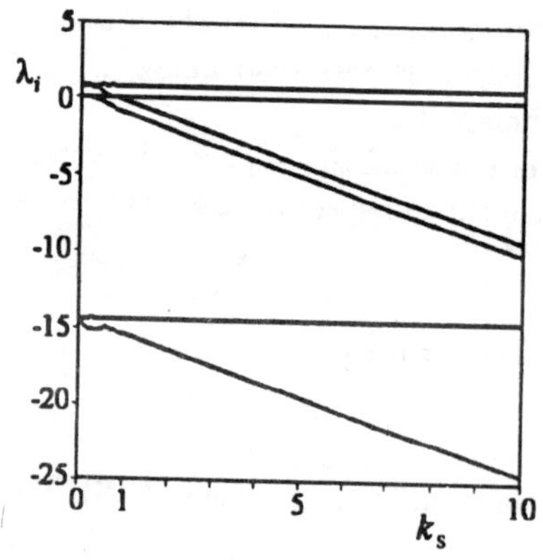

Fig. 7.18. Values of Lyapunov exponents of two coupled Lorenz systems versus k. In this example $n = \epsilon$, and we have assumed that $k_1 = k_2 = k_3 = k$

and introducing $\lambda_i(x)$, the Lyapunov exponents associated with the state difference, we have:

$$\sum_{i=1}^{n} \lambda_i(x) = \sum_{i=1}^{n} \lambda_i(x_1) - \sum_{i=1}^{n} k_i. \tag{7.22}$$

Full synchronisation takes place when the fixed point of the state difference is stable, i.e. $x = X_s = 0$. This stability forces fulfilling of the inequality:

$$\lambda_i(x)_{\max} < 0. \tag{7.23}$$

Hence, assuming $k_i = k$, for all i, *full synchronisation* of (7.16) is guaranteed by:

$$k > \lambda_i(x_1)_{\max}, \tag{7.24}$$

and, for the simple form of coupling we have assumed in system (7.15), it is clear that $\lambda_i(x) = k_i$ for all i, and hence the synchronisation value of the coupling parameter is given by:

$$k_s = \lambda_{\max}, \tag{7.25}$$

and it provides a measure for $\lambda_{\max}$, where $\lambda_{\max}$ represents the maximum Lyapunov exponents.

We can apply this property to determine the value of the largest Lyapunov exponent of the system (7.26):

$$\begin{aligned} \dot{y} &= z \\ \dot{z} &= q\cos(\omega t) + ay(1-y^2) - hz \\ &\quad -\epsilon N\left(\frac{\mu_0-\mu_1}{1+\lambda_1|z|} + \mu_0 + \lambda_2 z^2\right)\mathrm{sign}(z). \end{aligned} \tag{7.26}$$

The system (7.26) has been created by adding to the classical Duffing equation a term describing dry friction according to Popp–Stelter formula [7.26]. Its discontinuous nature makes any direct calculation of Lyapunov exponents virtually impossible.

Here, determination of the largest Lyapunov exponent has been done by coupling two identical systems (7.26) according to (7.15). After putting the considered system (7.26) in (7.15) the augmented system is as follows:

$$\begin{aligned} \dot{y}_1 &= z_1 \\ \dot{z}_1 &= q\cos(\omega t) + ay_1(1-y_1^2) - hz_1 \\ &\quad -\epsilon N\left(\frac{\mu_0-\mu_1}{1+\lambda_1|z_1|} + \mu_0 + \lambda_2 z_1^2\right)\mathrm{sign}(z_1) \\ \dot{y}_2 &= z_2 + k(y_1 - y_2) \\ \dot{z}_2 &= q\cos(\omega t) + ay_2(1-y_2^2) - hz_2 \\ &\quad -\epsilon N\left(\frac{\mu_0-\mu_1}{1+\lambda_1|z_2|} + \mu_0 + \lambda_2 z_2^2\right)\mathrm{sign}(z_2) + k(z_1 - z_2). \end{aligned} \tag{7.27}$$

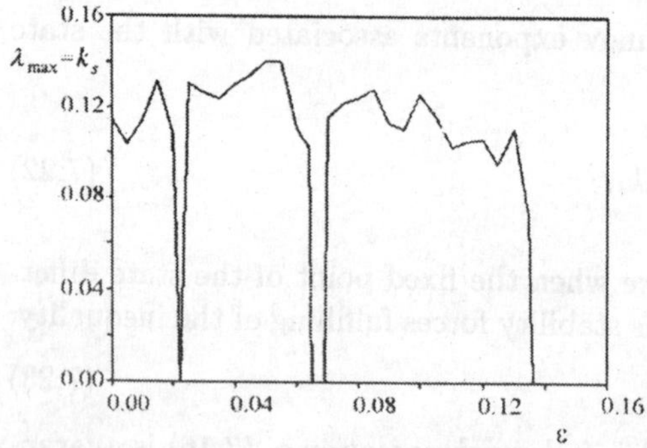

Fig. 7.19. Largest Lyapunov exponent versus ϵ parameter value for the investigated system

In our numerical simulations we have assumed the following values of parameters: $a = 1, h = 0.25, q = 0.3, \omega = 1.0, \lambda_1 = 1.42, \lambda_2 = 0.005, \mu_0 = 0.25, \mu_1 = 0.05, N = 1$.

Then the synchronisation value for these subsystems was determined, and this, according to (7.25) is equal to the largest Lypunov exponent of the system (7.26). This way a spectrum of Lyapunov exponents as a function of ϵ (representing the dry friction force) of the investigated system was obtained. The spectrum is presented in Fig. 7.19.

The bifurcation diagram of the system (7.26), shown in Fig. 7.20 as the dependence of z on ϵ parameter values, can be regarded as verification of the correctness of results: The ϵ parameter range is identical to that in Fig. 7.19.

The values of λ_{max} are obtained from bifurcation diagrams of *state difference* between subsystems (7.27) as a function of coupling parameter k for varying values of ϵ. An example of such a diagram is presented in Fig. 7.21.

We can see bands where chaotic motion occurs (Fig. 7.20) when the Lyapunov exponents take positive values (Fig. 7.19). Also periodic motion occurs when a zero value of the Lyapunov exponent is found in Fig. 7.19. These results support our hypothesis that the values of the largest Lyapunov expo-

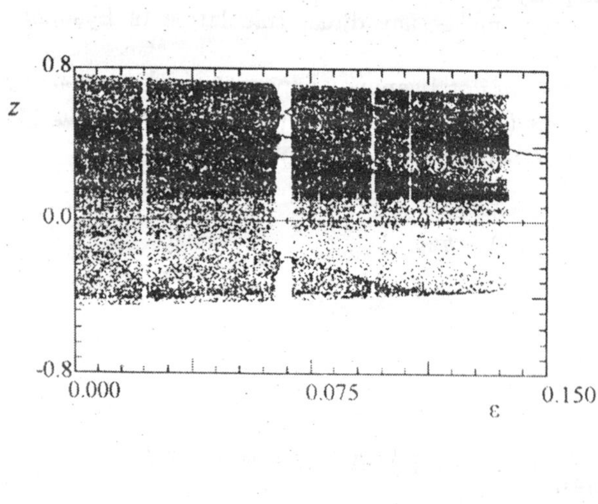

Fig. 7.20. Detail from the bifurcation diagram of z versus ϵ for Duffing's equation with friction (7.26)

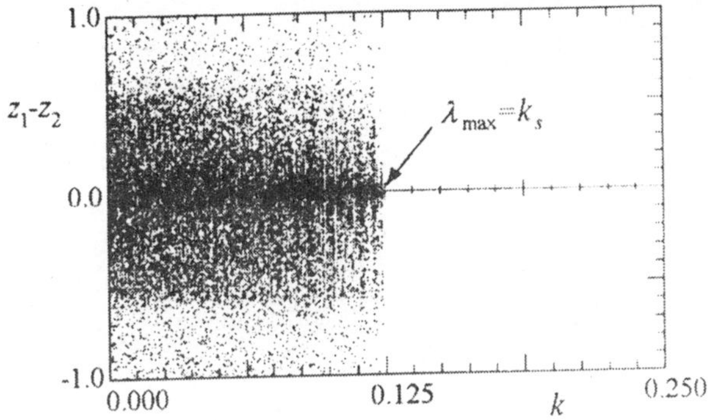

Fig. 7.21. Bifurcation diagram showing *state difference* versus k for coupled Duffing equations with dry friction (7.27); $\epsilon = 0.1$

nents obtained as above for the system with dry friction can agree with the real ones (impossible to determine by other methods) to a good approximation.

References

Chapter 1

1.1 Hayashi, C. (1964): Nonlinear Oscillations in Physical Systems, Mc. Graw-Hill, New York
1.2 Nayfeh, A.H. and Mook, D.T. (1979): Non-Linear Oscillations, J. Wiley and Sons, Chichester
1.3 Thomsen, J.J. (1997): Vibrations and Stability, Mc. Graw-Hill, London
1.4 Chen, Y., Leung, A.Y.T. (1998): Bifurcation and Chaos in Engineering, Springer, New York
1.5 Blekhman, I.I. (1999): Vibrational Mechanics, World Scientific, Singapore

Chapter 2

2.1 Hale, J. (1969): Ordinary differential equations, Wiley, (1969), New York
2.2 Carr, J. (1981): Applications of Center Manifold Theory, Springer, New York
2.3 Shilnikov, L.P., Shilnikov, A.L., Turaev, D.V., Chua, L.O. (1998): Methods of Qualitative Theory in Nonlinear Dynamics, World Scientific, Singapore
2.4 Wolf, A., Swift, J., Swinney, H. and Vastano, A. (1985): Determining Lyapunov exponents from time series, Physica, **16D**, 285–314
2.5 Parker, T., Chua, L. O. (1989): Practical Numerical Algorithms for Chaotic Systems, Springer, New York
2.6 Kapitaniak, T. (1991): Chaotic Oscillations in Mechanical Systems, Manchester University Press, Manchester
2.7 Kaplan, W. (1973): Advanced Calculus, Addison-Wesley, Reading MA
2.8 Whitney, H. (1936): Differentiable manifolds, Ann. Math, **37**, 645-708
2.9 Takens, F. (1981): Detecting strange attractors in turbulence, In: Dynamical Systems and Turbulence - Warwick 1980, Lecture Notes in Mathematics, **898**, Rand, D.A., and Young, L.-S. (eds), Springer, New York

Chapter 3

3.1 May, R. (1976): Simple Mathematical models with very complicated dynamics, Nature, **26**, 459-467
3.2 Collet, P., Eckmann, J. P. (1980): Iterated Maps of the Interval as Dynamical Systems, Birkhäuser, Boston
3.3 Feigenbaum, M. (1978): Qualitative universality for a chaos of nonlinear transformations, J. Stat. Phys., **19**, 5-32
3.4 Lorenz, E. (1963): Deterministic nonperiodic flow, J. Atmos. Sci., **2**, 130-141

Chapter 4

4.1 Edgar, G.A. (1990): Measure, Topology and Fractal Geometry, Springer, New York
4.2 Hausdorff, F. (1919): Dimension und ausseres Mass, Mathematische Annalen, **79**, 157-179
4.3 Ott, E. (1993): Chaos in Dynamical Systems, Cambridge University Press, Cambridge
4.4 McDonald, S.W., Grebogi, C., Ott, E. and Yorke, J.A. (1985): Fractal basin boundaries, Physica, **17D**, 125-149
4.5 Kaplan, J.L., Yorke, J.A. (1979): Chaotic behaviour of multidimensional difference equations, In: Functional Differential Equations and Approximations of Fixed Points, H.- O. Peitgen and H.- O. Walter, Lecture Notes in Mathematics, **730**, Springer, Berlin
4.6 Mandelbrot, B. (1982): The Fractal Geometry of Nature, Freeman, San Francisco
4.7 Kolmogorov, A.N. (1958): A new metric invariant of transitive dynamical systems, Dokl. Akad. Nauk SSSR, **119**, 861-918
4.8 Grassberger, P., Procacia, J. (1983): Measuring the strangeness of strange attractors, Physica, **9D**, 189-204
4.9 Smale, S. (1967): Differentiable Dynamical Systems, Bull. Amer. Math. Soc., **73**, 747-774
4.10 Abraham, R.H., Show, C.D. (1984): Dynamics - The geometry of behaviour, Part III, Ariel Press, Santa Cruz

Chapter 5

5.1 Newhouse, S., Ruelle, D. and Takens, F. (1978): Occurrence of strange axiom attractors near quasiperiodic flows on T^m, $m > 3$, Commun. Math. Phys., **64**, 35-40
5.2 Landau, L. D. (1944): On the problem of turbulence, C. R. Acad. Sci. URSS, **44**, 311-318
5.3 Pomeau, Y., Manneville, P. (1980): Intermittent transition to turbulence in dissipative dynamical systems, Commun Math. Phys., **74**, 189-197
5.4 Sato, M., Sano, M. and Sawada, Y. (1983): Universal scaling property in bifurcation structure of Duffing's and generalized Duffing's equations, Phys. Rev., **82A**, 1654- 1658
5.5 Kapitaniak, T. (1991): Chaotic Oscillations in Mechanical Systems, Manchester University Press, Manchester
5.6 Ueda, Y. (1979): Randomly transitional phenomena in the system governed by Duffing's equation, J. Stat. Phys., **20**, 181-196

Chapter 6

6.1 Andronov, A.A., Vitt, E.A., Khaiken, S.E. (1966): Theory of oscillations, Pergamon Press, Oxford
6.2 Parlitz, V., Lauterborn, W. (1987): Period-doubling cascades and devil's staircase of the driven van der Pol oscillator, Phys. Rev., **36A**, 1428-1434

6.3 Wiercigroch, M. (1998): Chaotic vibration of a simple model of the machine tool-cutting process system, J. Vibr. Acoustics, **119**, 468-475
6.4 Blazejczyk-Okolewska, B., Czolczynski, K., Kapitaniak, T., Wojewoda, J. (1999): Chaotic Mechanics in Systems with Impacts and Friction, World Scientific, Singapore
6.5 Wiercigroch, M., de Kraker (1999): Nonlinear Dynamics and Chaos in Mechanical Systems with Discontinuities, World Scientific, Singapore
6.6 Kunick, A., Steeb, W.-H. (1987): Chaos in systems with limit cycle, Int. J. Nonlinear Mech., **22**, 349-362
6.7 Swinney, H. L. (1983): Observations of order and chaos in nonlinear systems, Physica, **7D**, 3-15
6.8 Rapp, P. E. (1986): Oscillations and chaos in cellular metabolism and physiological systems, (Chaos, Holden, A.V., ed.), Princeton University Press, Princeton
6.9 Gray, P., Scott, S. (1986): Oscillations in forced chemical reactions, Ber. Bunsenges Phys. Chem., **90**, 985-997
6.10 Sel'kov, E. E. (1968): Self-oscillations in glycolysis, Part. 1. A simple kinetic models, European J. Biochem., **4**, 79-98
6.11 Merkin, J., Needham, D., Scott, S. (1986): Oscillatory chemical reactions in closed venels, Proc. R. Soc. Lond., **A406**, 299-318
6.12 El Naschie, M.S., Al Athel, S. (1989): On the connection between statical and dynamical chaos, Z. Natuf., **44a**, 645-652
6.13 El Naschie, M.S. (1988): Generalized bifurcation and shell buckling as spatial statical chaos, ZAMM, T367-377
6.14 El Naschie, M.S. (1990): Stress stability and chaos, Mc Graw-Hill, London
6.15 Kapitaniak, T. (1991): Chaotic Oscillations in Mechanical Systems, Manchester University Press, Manchester
6.16 Ogorzalek, M. (1997): Chaos and Complexity in Nonlinear Electronic Circuits, World Scientific, Singapore
6.17 Chua, L.O. (1992): The genesis of Chua's circuit, Archiv. fur Elektronik und Ubertragungstechnik, **46**, 250-257
6.18 Madan, R. (1993): Chua's circuit: paradigm for chaos, World Scientific, Singapore
6.19 Matsumoto, T. (1984): A chaotic attractor from Chua's circuit, IEEE Trans. Circuits Syst., **CAS-31**, 1055-1058
6.20 Kocarev, L. (1999): Introduction to Nonlinear Dynamics: Chua's Circuit, World Scientific, Singapore
6.21 Kapitaniak, T. (1990): Chaos in Systems with Noise, World Scientific, Singapore
6.22 Valis, G.K. (1988): Conceptual models of El Nino and the Southern Oscillation, J. Geophys. Res., **93**, 13979-14019
6.23 Weickmann, K.M., (1991): El. Nino/Southern Oscillation and Madden Julian oscillations during 1981-82, J. Geophys. Res., **96 Supp**, 3187-3199

Chapter 7

7.1 Ott, E., Grebogi, C., Yorke, J.A. (1990): Controlling chaos, Phys. Rev. Lett., **64**, 1196-1199
7.2 Romeiras, F., Ott, E., Grebogi, C., Dayawansa, W.P. (1992): Controlling chaotic dynamical system, Physica, **58D**, 165-180

7.3 Ogorzalek, M. (1994): Chaos control: How to avoid chaos or take advantage of it, J. Franklin Inst., **331B**, 681-704
7.4 Ditto, W.L., Rauseo, S.W., Spano, M.L. (1991): Experimental control of chaos, Phys. Rev. Lett., **65**, 3211-3216
7.5 Hunt, E.H. (1991): Stabilizing high-periodic orbits in a chaotic system: the diode resonator, Phys.Rev. Lett., **67**, 1953-1959
7.6 Mehta, N., Henderson, R. (191): Controlling chaos to generate aperiodic orbit, Phys. Rev., **44A**, 4861-4869
7.7 Tel, T. (1991): Controlling transient chaos, J. Phys. A., **24**, 11359-11365
7.8 Dressler, V., Nitsche, G. (1992): Controlling chaos using time delay coordinates, Phys. Rev. Lett., **68**, 1-5
7.9 Singer, J., Wang, Y.-Z., Bau, H.H. (1991): Controlling a chaotic system, Phys. Rev. Lett., **66**, 1123-1126
7.10 Pyragas, K. (1992): Continuous control of chaos by self-controlling feedback, Phys. Lett., **170A**, 421-427
7.11 Ueda, Y. (1979): Randomly transitional phenomena in the system governed by Duffing's equation, J. Stat. Phys., **20**, 181-196
7.12 Kapitaniak, T. (1991): Chaotic Oscillations in Mechanical Systems, Manchester University Press, Manchester
7.13 Wolf, A., Swift, J., Swinney, H. and Vastano, A. (1985): Determining Lyapunov exponents from time series, Physica, **16D** 285-314
7.14 Kapitaniak, T. (1996): Controlling Chaos, Academic Press, London
7.15 Chen, G., Dong, X. (1998): From Chaos to Order, World Scientific, Singapore
7.16 Schuster, H.G. (1999): Handbook of Chaos Control, J. Wiley and Sons, Chichester
7.17 Pecora, L., Carroll, T. (1990): Synchronization in chaotic systems, Phys. Rev. Lett., **64**, 821- 825
7.18 Chua, L.O., Kocarev, L., Eckert, K. and Itoh, M. (1992): Experimental chaos synchronization in Chua's circuit, Int. J. Bif. Chaos, **2**, 705-712
7.19 Kapitaniak, T. Chua, L.O., Zhong, G.Q. (1994): Experimental synchronization of chaos using continuous control, Int. J. Bif. Chaos, **4**, 483-488
7.20 Qu, Z., Hu, G., Ma, B. (1993): Note on continuous chaos control, Phys., Lett., **178A**, 265-272
7.21 Cuomo, K., Oppenheim, A. (1993): Circuit implementation of synchronized chaos with applications to communication, Phys. Rev. Lett., **71**, 65-69
7.22 Kocarev, L. Halle, K., Eckert, K., Chua, L.O. (1992): Experimental demonstration of secure communications via chaos synchronization, Int. J. Bif. Chaos, **2**, 709-716
7.23 Halle, K., Wu, C.W., Itoh, M. and Chua L.O. (1993): Spread spectrum communication though modulation of chaos, Int. J. Bifurcat. Chaos, **3**, 469-477
7.24 Parlitz, V., Lauterborn, W. (1987): Period-doubling cascades and devil's staircase of the driven van der Pol oscillator, Phys. Rev., **36A**, 1428-1434
7.25 Perez, G. and Cerdeira, H. (1995): Extracting messages masked by chaos, Phys. Rev. Lett. **74**, 1970-1973
7.26 Popp, K. and Stelter, P. (1990): Nonlinear oscillations of structures induced by dry friction, in: *Non-linear Dynamics in Engineering Systems*, ed. W. Schiehlen, Springer, New York

Index